VARIATIONAL PROBLEMS IN TOPOLOGY

VARIATIONAL PROBLEMS IN TOPOLOGY

THE GEOMETRY OF LENGTH, AREA AND VOLUME

A. T. FOMENKO

Faculty of Mathematics and Mechanics,
Moscow State University, USSR

CRC Press
Taylor & Francis Group
Boca Raton London New York

CRC Press is an imprint of the
Taylor & Francis Group, an **informa** business

Originally published as топологические вариационные задачи
by Moscow University Press, Moscow

First published 1990 by Gordon and Breach Science Publishers

Published 2019 by CRC Press
Taylor & Francis Group
6000 Broken Sound Parkway NW, Suite 300
Boca Raton, FL 33487-2742

© 1990 by Taylor & Francis Group, LLC
© 1984 by Moscow University Press, Moscow
CRC Press is an imprint of Taylor & Francis Group, an Informa business

First issued in paperback 2019

No claim to original U.S. Government works

ISBN 13: 978-0-367-45603-0 (pbk)
ISBN 13: 978-2-88124-740-8 (hbk)

Visit the Taylor & Francis Web site at
http://www.taylorandfrancis.com

and the CRC Press Web site at
http://www.crcpress.com

Library of Congress Cataloging-in-Publication Data
Fomenko, A. T.
    [Topologicheskie variafsionnye zadachi. English]
    Variational problems in topology / A.T. Fomenko
        p.   cm.
    Translation of: Topologicheskie variafsionnye zadachi.
    Includes bibliographical references.
    ISBN 2-88124-740-7 (Switzerland)
    1. Topology. 2. Variational inequalities (Mathematics)
I. Title.
QA611.F6313   1990
514--dc20   89-77326
                                                             CIP

CONTENTS

PREFACE

Within the framework of modern geometry, there stand out problems bordering on several branches of science: topology, variational calculus, algebra and the theory of differential equations. In this book, we consider some important variational problems in topology (both solved and unsolved) and apply modern analytical and topological methods as a means of research.

Three interrelated topics are dealt with in this book. Chapter I has more of a teaching-aid character, and is dedicated to some basic concepts of topology, *vis.* homology and cohomology, the theory of obstruction to the extension of a continuous mapping, fibrations and their simplest properties. Chapter I is not self-sufficient, since it is intended as an auxiliary for other chapters in which the above topological concepts play an important role. With reference to this, the presentation of the auxiliary topological material is simplified as much as possible; awkward technical and formal discussions are replaced by precise references to corresponding sources, which enables the author to concentrate on revealing visual geometric mechanisms underlying the problems under study.

The common feature of the second range of questions studied in the present book is the important, sometimes essential, role played by Morse functions on smooth manifolds, i.e. functions whose critical points are non-degenerate. Much attention is paid not to classical Morse theory, which is well expounded in textbooks, but to more modern aspects of the topology of manifolds in relation to Morse functions. Chapter II, in particular, touches upon questions on Morse theory of many-valued functions, useful for solving some problems in modern mathematical physics. One of the important topological problems in which Morse functions play a significant role is the description of manifolds of small dimension. In Chapter III, we try (as far as is possible in a survey) to acquaint the reader with questions pertaining to the famous Poincaré problem (whether a 1-connected closed three-dimensional manifold is homeomorphic to the standard three-dimensional sphere) and its multidimensional analogs. We expand in particular a simple new recognition algorithm of the standard three-dimensional sphere in the class of Heegard diagrams of genus two, based on the so-called "separating vertex". In conclusion, we outline the solution of the four-dimensional Poincaré problem obtained by Freedman in 1981. The presentation of Point 2 in Section 1 and Point 3 in Section 3 is based on the material kindly placed at our disposal by S. V. Matveev, to whom the author would like to express his sincere gratitude.

The third topic elaborated in Chapter IV is most synthetic of all. Here, we discuss minimal surfaces in the wide sense, and consider multidimensional minimal surfaces, extremals of the Dirichlet functional, stratified extremals of stratified functionals, and integral currents, as well as classical minimal surfaces realizing locally minimum areas. Various topological problems related to these concepts and their modern applications use, in one form or another, topological facts of various levels of complexity, given in Chapter I. The reader should note the following feature of this section: many deep and non-trivial geometric statements of modern variational calculus can be demonstrated on the level of real physical experiments that date back as far as the 19th century and are still important. We discuss, in particular, some of Plateau's experiments with soap films, making wide use of graphic representation, which enables us to display necessary physical effects in a visual and compact form.

Since this book is intended for the general reader, all basic facts are reduced to a minimum indispensable for comprehension of the material discussed. However, knowledge of the basic notions of geometry, set forth in standard textbooks, e.g. in References 1, 4, 5 and 6, will undoubtedly be useful for quicker penetration into the essence of the questions under discussion.

A. T. Fomenko

CHAPTER I/PRELIMINARIES

1. Singular and Cellular Homology Groups

1. Singular Chains and Homology Groups

Recent results in the theory of variational topological problems show the importance of homology groups. Let us recall some necessary facts and certain algebraic and topological constructions which are used in the sequel.

Consider a Euclidean space $\mathbf{R}^{k+1}$ with respect to Cartesian co-ordinates $x_1, \ldots, x_{k+1}$ and the standard simplex Δ^k given by $x_1 + \ldots + x_{k+1} = 1$, $x_1 \geqslant 0, \ldots, x_{k+1} \geqslant 0$, of dimension k.

DEFINITION 1. A continuous mapping f of the standard simplex Δ^k into a topological space X is called *a singular k-dimensional simplex f^k of X*. A formal linear combination of singular simplices f^k of X with integer coefficients, only finitely many of them being different from zero, is called *an integer k-dimensional singular chain $c = \sum_i \alpha_i f_i^k$ of X*. The set of all k-dimensional chains is obviously transformed into an Abelian additive group $C_k(X)$ referred to as *the group of k-dimensional chains* of the space X. The boundary operator $\partial_k : C_k(X) \to C_{k-1}(X)$ is the homomorphism defined on the generators as

$$\partial_k f^k = \sum_{i=0}^{k} (-1)^i f_i^{k-1},$$

where f_i^{k-1} is the restriction of the mapping f^k to the i-th face of the simplex Δ^k.

It is evident that $\partial_k \partial_{k+1} \equiv 0$; therefore, $\operatorname{Ker} \partial_k \supset \operatorname{Im} \partial_{k+1}$.

DEFINITION 2. The quotient group $\operatorname{Ker} \partial_k / \operatorname{Im} \partial_{k+1}$ is called *the k-dimensional singular homology group $H_k(X)$ of X*. The elements of $\operatorname{Ker} \partial_k$ are called *cycles*; those of $\operatorname{Im} \partial_{k+1}$, *boundaries*.

One important property easily follows from this definition, viz., singular homology groups are both topologically and homotopy invariant, i.e., they do not alter under a space homeomorphism or if the space is replaced by a homotopy-equivalent one. The collection of groups $C_k(X)$ and connecting homomorphisms ∂_k is naturally organized into the sequence

$$\ldots \to C_k \xrightarrow{\partial_k} C_{k-1} \to \ldots \to C_1 \xrightarrow{\partial_1} C_0 \xrightarrow{\varepsilon} \mathbf{Z} \to 0,$$

where $\partial_k \partial_{k+1} \equiv 0$ and ε is an epimorphism. This sequence is called a *chain complex*. A mapping commuting with the homomorphisms ∂_k of one chain complex into another is said to be a chain mapping. We will say that given is a chain homotopy D of a complex C into a complex C', connecting two mappings φ and ψ of C, where $\varphi = \{\varphi_k : C_k \to C_k'\}$, $\psi = \{\psi_k : C_k \to C_k'\}$, if there is a

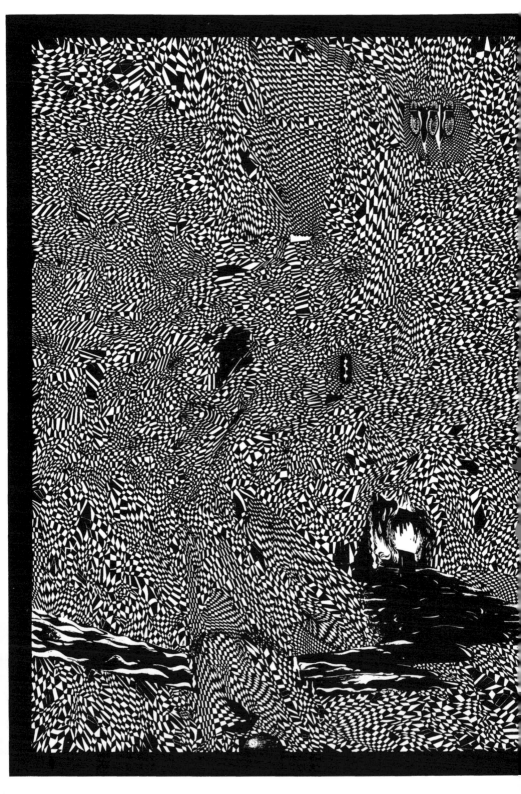

collection of such homomorphisms $D_k : C_k \to C'_{k+1}$ that, for each k, $D_{k-1}\partial_k + \partial'_{k+1}D_k = \varphi_k - \psi_k$. Two chain mappings φ and ψ connected by a chain homotopy are sometimes said to be *chain-homotopic*. It follows from the definition of homology groups that chain-homotopic mappings induce identical mappings of homology groups $H_k(C)$ *into* $H_k(C')$, where $H_k(C) = \mathrm{Ker}\, \partial_k / \mathrm{Im}\, \partial_{k+1}$.

To calculate homology, the so-called exact sequence of a pair can be extremely useful. Let X and Y be two topological spaces, where Y is a closed subspace of X. It is clear that $C_k(Y) \subset C_k(X)$, and we may consider the group of relative chains $C_k(X, Y) = C_k(X)/C_k(Y))$. Since the boundary operator acts as follows, viz., $\partial_k : C_k(X) \to C_{k-1}(X)$, $\partial_k : C_k(Y) \to C_{k-1}(Y)$, it induces a certain operator $C_k(X, Y) \to C_{k-1}(X, Y)$ which for simplicity we denote by the same symbol ∂_k. We can define the groups $\mathrm{Ker}\, \partial_k = Z_k(X, Y) \supset B_k(X, Y) = \mathrm{Im}\, \partial_k$ (relative cycles and relative boundaries), which enables us to consider the quotient group $H_k(X, Y) = Z_k(X, Y)/B_k(X, Y)$ called the group of relative k-dimensional homology groups of the space X modulo the subspace Y. It is easy to verify that for relative homology groups both topological and homotopy invariance hold. We now pass to the construction of a new operator $\partial : H_k(X, Y) \to H_{k-1}(Y)$. Let $z_k \in C_k(X, Y)$ be a relative cycle, and $\tilde{z}_k \in C_k(X)$ its arbitrary representative in the coset. Since $\partial_k z_k = 0$, $\partial_k \tilde{z}_k \in C_{k-1}(Y)$. Denote this absolute cycle by $\partial z_k \in Z_{k-1}(Y)$. Its homology class does not depend on the choice of a homology class representative of z_k. Hence, we define a certain homomorphism (operator) $\partial : H_k(X, Y) \to H_{k-1}(Y)$ which we also denote by ∂, and call boundary (Fig. 1). Further, denote the embedding by $i : Y \to X$; it then induces the homomorphism $i_* : H_k(Y) \to H_k(X)$. Since any absolute cycle can be regarded as relative (modulo the subspace Y), one more natural mapping $j_* : H_k(X) \to H_k(X, Y)$ is given rise.

THEOREM 1. *The following sequence of groups and homomorphisms is exact, i.e., the image of each incoming homomorphisms coincides with the kernel of each outgoing, viz.,*

$$\ldots \to H_{k+1}(X, Y) \overset{\partial}{\to} H_k(Y) \overset{i_*}{\to} H_k(X) \overset{i_*}{\to} H_k(X, Y) \overset{\partial}{\to} H_{k-1}(Y) \to \ldots$$

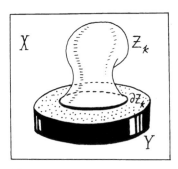

Figure 1

PROOF. LET US VERIFY, E.G., THE EXACTNESS IN THE TERM $H_k(X, Y)$. If $\partial\alpha = 0$, $\alpha \in H_k(X, Y)$, then ∂z_k is homologous to zero in Y, where z_k is a representative of α, i.e., $z_k \in Z_k(X, Y)$. But then, adding to z_k a k-dimensional chain realizing this homology in the subspace Y, we obtain a k-dimensional chain, a cycle now from the standpoint of the ambient space, i.e., we have represented the original element as the image of a certain element β, i.e., $\alpha = j_* \beta$, $\beta \in H_k(X)$ (Fig. 2). Thus, $\operatorname{Ker} \partial \subset \operatorname{Im} j_*$. The inverse inclusion $\operatorname{Ker} \partial \supset \operatorname{Im} j_*$ follows from the fact that any absolute cycle in X, which is then regarded as relative, has the null boundary in Y. The exactness of the sequence in the other terms is verified similarly.

The following simple properties of exact sequences will be used in the sequel (we leave the proof to the reader as an exercise):
  (i) The sequence $0 \to A \to 0$ is exact if and only if $A = 0$.
  (ii) The sequence $0 \to A \overset{\alpha}{\to} B \to 0$ is exact if and only if the groups A and B are isomorphic, and the homomorphism α is an isomorphism.
  (iii) The sequence $0 \to A \overset{i}{\to} B \overset{\pi}{\to} C \to 0$ is exact if and only if the group A is a subgroup of B, the homomorphism $i: A \to B$ being an inclusion (monomorphism), $C = B/A$ (quotient group) and $\pi: B \to B/A = C$, the natural projection onto the quotient group).
It turns out that there is a reduction of relative homology groups to the absolute.

2. Cell Complexes and Barycentric Subdivisions

If X, Y are path-connected, then $H_0(X, Y) = 0$. A topological space X is called a *cell complex* if it is represented in the form of the union of disjoint subsets σ^k known as cells, the closure of each being the image of a closed k-dimensional disk D^k under some continuous mapping called characteristic (which is a homeomorphism on its interior). Further, the boundary of each cell, i.e., the disk boundary image under the characteristic mapping, must be contained in the union of finitely many cells of smaller dimensions. Finally, it is required

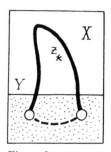

Figure 2

that the subset of X should be closed if and only if all full inverse images of the subset intersections with all cells are closed under characteristic mappings.

A complex is said to be finite if it consists of a finite number of cells. The union of cells of dimensions not exceeding k is called the k-dimensional skeleton of the complex. We will call a pair of spaces (X, Y) a cell pair if X and Y are cell complexes and the closed subspace Y is a cell subcomplex of X. We shall confine ourselves to finite complexes. Let X/Y be a space obtained from X by identifying a closed subspace Y with a point.

THEOREM 2. *Let (X, Y) be a cell pair. Then $H_k(X, Y) = H_k(X/Y)$ for $k \neq 0$.*

PROOF. Let CY be the cone over Y, i.e., the space obtained from the cylinder $Y \times I$ by contracting the upper base to a point (Fig. 3). Construct a new space $X \cup CY$, i.e., identify the subspace Y of X with its base in CY (Fig. 4). Since X, Y are finite complexes, contracting CY on itself to a point, we obtain a homotopy equivalence $X \cup CY \approx X/Y$ (Fig. 5). Thus, to prove the theorem, it suffices to establish the relations $H_k(X, Y) = H_k(X \cup CY)$ for $k > 0$. It follows from the exact sequence of the pair $(X \cup CY, *)$, where $*$ is a point (cone vertex)

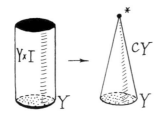

Figure 3

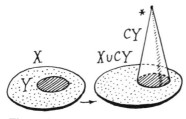

Figure 4

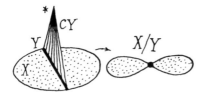

Figure 5

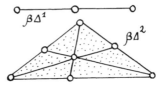

Figure 6

that $H_k(X \cup CY, *) = H_k(X \cup CY)$ for $k > 0$; therefore, we must prove that $H_k(X, Y) = H_k(X \cup CY, *)$ for $k > 0$. Before passing to the proof, we dwell on the so-called barycentric subdivision.

Let Δ^k be the standard simplex. Its barycentric subdivision $\beta\Delta^k$ is defined by induction. If $k = 1$, then $\beta\Delta^1$ is obtained by introducing a new vertex at the midpoint of the segment Δ^1. If $k = 2$, then $\beta\Delta^2$ is obtained by joining the triangle centroid to the vertices and midpoints (Fig. 6). Finally, for an arbitrary k, the subdivision $\beta\Delta^k$ is obtained a follows: Distinguish the center of Δ^k, and break the simplex Δ^k into pyramids with vertices at this center and bases which are the simplices of the barycentric subdivision of the boundary of Δ^k. Now, let $f: \Delta^k \to X$ be an arbitrary singular simplex of the space X. Denote by βf the chain equal to the sum of all singular simplices obtained by restricting the original mapping f to the k-dimensional simplices of the barycentric subdivision $\beta\Delta^k$ of the original simplex Δ^k. Associating each singular simplex f, i.e., an elementary chain in $C_k(X)$, with the singular chain βf, we obtain a homomorphism $\beta_k: C_k(X) \to C_k(X)$. We state that the collection of mappings $\beta = \{\beta_k\}$ determines a chain mapping β of the chain complex $C(X)$ into itself, which is chain-homotopic to the identity. The fact that the mapping β is chain stems from the restriction of the singular simplex formal sum to the barycentric subdivision being equal to the formal sum of these restrictions. Construct explicitly a chain homotopy D connecting β with the identity mapping. For each $k > 0$, we define a partition of the prism $\Delta^k \times I$ into the sum of simplices in the following way: When $k = 0, 1$ and 2, the partition is shown in Fig. 7. Now, we describe the induction process of the partition $\Delta^k \times I$: If the partition has already been defined for $q < k$, then, on part of the boundary $\Delta^k \times I$, viz., the union ("cup") $(\Delta^k \times 0) \cup (\partial\Delta^k \times I)$, it should be constructed so as to make $\Delta^k \times 0$ the standard simplex, and decompose $\partial\Delta^k \times I$ by the induction process when $q < k$. Further, decompose the prism into k-dimensional simplices whose bases are $(k-1)$-dimensional simplices of the above partition, and the vertex is the centre of the upper face. It is clear that the upper face is subject to the barycentric subdivision.

Now, let $f: \Delta^k \to X$ be a singular simplex. By $D_k f \in C_{k+1}(X)$, we denote the $(k+1)$-dimensional chain which is the sum of all $(k+1)$-dimensional singular simplices obtained by restricting the mapping $\Phi: \Delta^k \times I \to X$, where $\Phi(x,t) = f(x)$, to the simplices of the constructed partition of the cylinder $\Delta^k \times I$. We obtain homomorphisms $D_k: C_k(X) \to C_{k+1}(X)$ which determine the

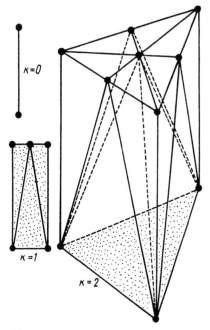

$K=0$

$K=1$

$K=2$

Figure 7

required chain homotopy connecting the identity mapping of the complex $C(X)$ onto itself to the barycentric mapping. The proof that $\{D_k\}$ determine a chain homotopy is left to the reader. We now turn to the proof of the theorem. Consider an embedding mapping for the pairs $(X,Y) \to (X \cup CY, CY)$, inducing the homomorphism $\alpha : H_k(X,Y) \to H_k(X \cup CY, CY) = H_k(X \cup CY, *)$. Here, we have made use of the fact that $CY \approx *$, i.e., the cone is continuously contractible to a point. Prove that α is an epimorphism. Let $z \in Z_k(X \bigcup CY, CY)$ be a certain cycle. In the group $Z_k(X,Y)$, we should find a cycle carried by the given mapping into another cycle, homologous to z in $Z_k(X \cup CY, CY)$. Represent CY as the union of two subsets, viz., the cone A, part of CY, consisting of the points for which $t \geqslant 1/2$ and the frustum of the cone B, part of CY, consisting of the points for which $t \leqslant 1/2$ (Fig. 8). Refining singular simplices composing the cycle z by barycentric subdivisions, we always obtain cycles homologous to it, which consist of smaller and smaller simplices. Since the number of simplices is finite and X, Y is compact, there exists such a sufficiently fine barycentric subdivision that if some singular simplex in it intersects A, then the simplex lies in CY wholly. Here, we consider the image of the standard simplex under a mapping determining a singular simplex. This cycle z' is homologous to z. Remove from z' all simplices intersecting A. Since this operation is performed within CY, from the standpoint of relative homology, a new cycle z'' is in the same homology class as z', and therefore, as

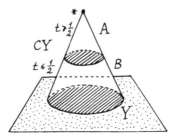

Figure 8

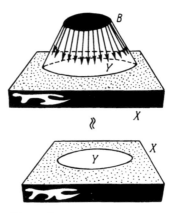

Figure 9

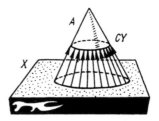

Figure 10

z. Thus, $z'' \in H_k(X \cup B, B)$. By the homotopy invariance of homology groups, we obtain $H_k(X \cup B, B) = H_k(X, Y)$, since $(X \cup B, B) \approx (X, Y)$. (Fig. 9). We have thus explicitly exhibited a certain cycle $z'' \in H_k(X, Y)$ carried by the homomorphism α into the cycle $z \in H_k(X \cup CY, CY)$. The epimorphism of α is thereby proved. The proof of its monomorphism is similar, and we leave it to the reader (Fig. 10).

3. Cellular Homology and Computation of Singular Homology of the Sphere

Outlined above has been one of the methods for introducing homology groups with the use of singular simplices. However, for cell complexes, the so-called cellular homology may be defined, which turns out to coincide with singular homology and possess the important advantage of being much easier to compute. The calculation of singular homology even in the simplest case where the space consists of one point requires some, though elementary, reasoning. If the space is more complex, then the calculation of its singular homology sharply becomes more complicated. Therefore, it is cellular homology that is of an especially frequent use in practice. We introduce it on the basis of the already known concept of singular homology, which will help us not only prove the singular and cellular homology group coincidence, but exhibit a clear method for their computation.

Let us compute the singular homology of the sphere. Since the zero-dimensional sphere S^0 consists of a pair of points, we obtain $H_k(S^0)=0$ for $k \neq 0$ and $H_0(S^0)=\mathbf{Z} \oplus \mathbf{Z}$.

LEMMA 1. *The singular homology groups of the n-dimensional sphere S^n where $n>0$ are of the form $H_k(S^n)=0$ for $k \neq 0, n$, and $H_k(S^n)=\mathbf{Z}$ for $k=0, n$.*

The proof follows from considering the exact homology sequence of the pair (D^{n+1}, S^n) and from the fact that the disk is contractible to a point.

Recall the definition of the wedge of two topological spaces. Distinguish two points x_0 and y_0 in two spaces X and Y, respectively. We construct a new space $X \vee Y$, by identifying the points (Fig. 11). No other points are identified.

LEMMA 2. *Let $X= \vee S_i^n$ be the union of n-dimensional spheres S_i^n indexed by i, where $1 \leqslant i \leqslant N$. If $n>0$, $k>0$, then the isomorphism $H_k(X)= \bigoplus_i H_k(S_i^n)=\mathbf{Z} \oplus \ldots \oplus \mathbf{Z}$ (N times) exists.*

For proof, it suffices to consider an exact sequence of the pair $(\vee D_i^n, \vee \partial D_i^n)$, with $\vee D_i^n / \vee \partial D_i^n = \vee S_i^n$ (Fig. 12).

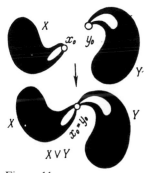

Figure 11

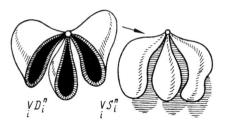

$$\bigvee_i D_i^n \qquad \bigvee_i S_i^n$$

Figure 12

Note the fact that will be of use in the sequel. As a generator in the group $Z = H_n(D^n, S^{n-1})$, we can take the homology class of the simplest singular chain $1 \cdot f$, where $f : \Delta^n \to D^n$ is a homeomorphism of the simplex onto the disk. Therefore, the orientation of a sphere may be given by fixing the generator in the group $Z = H_n(S^n)$. Change in the orientation is equivalent replacing the unit element 1 by the opposite.

Now, define the groups of cellular chains. Let X be a finite cell complex. Try computing its singular homology in terms of the cells and their characteristic mappings, i.e., in those terms in which the complex is given. Denote the set of all k-dimensional cells of X by X_k. Let X^k be the k-dimensional *skeleton* of X. We assume that the orientations of all cells are fixed. We then number all k-dimensional cells. Let A_k be an index set. Then, by Lemma 2,

$$H_i(X^k, X^{k-1}) = H_i(\bigvee_{\alpha \in A_k} S_\alpha^k) = \begin{cases} 0 & \text{when } i \neq k, \\ P_k(X) & \text{when } i = k \end{cases}. \text{ Here, a free Abelian group is}$$

denoted by $P_k(X)$, and its generators are in a one-to-one correspondence with A_k. Since the elements of this group are naturally identified with linear combinations of the form $\sum_\alpha a_\alpha \sigma_\alpha^k$, where σ_α^k are the k-cells of X, $P_k(X)$ is finitely generated. We will call $P_j(X)$ the group of cellular k-chains of X. The groups $P_k(X)$ and $C_k(X)$ are non-isomorphic in the general case. Before proceeding further, consider the so-called exact homology sequence of a triple, which is a version of a sequence of a pair. Let (X, Y, Z) be a triple of spaces, where Y and Z are closed in X. Consider two embeddings $(Y, Z) \to (X, Z)$ and $(X, Z) \to (X, Y)$. Let $\partial : H_k(X, Y) \to H_{k-1}(Y, Z)$ be the boundary homomorphism generated by the homomorphism $\partial : H_k(X, Y) \to H_{k-1}(Y)$ (see its definition above) due to the fact that each absolute cycle from $H_{k-1}(Y)$ may be regarded as relative mod Z, i.e., as an element of the group $H_{k-1}(Y, Z)$. We then obtain the sequence

$$\ldots \to H_k(X, Y) \xrightarrow{\partial} H_{k-1}(Y, Z) \to H_{k-1}(X, Z) \to H_{k-1}(X, Y) \xrightarrow{\partial} \ldots .$$

Verification of its exactness is left to the reader.

Returning to the groups $P_k(X) = H_k(X^k / X^{k-1})$ and $P_{k-1}(X) = H_{k-1}(X^{k-1} / X^{k-2})$, we can consider the exact sequence of the triple (X^k, X^{k-1}, X^{k-2}). At present, we only need the homomorphism $H_k(X^k, X^{k-1}) \to$

$H_{k-1}(X^{k-1}, X^{k-2})$, which is, in our new notation, designated by $P_k(X) \to P_{k-1}(X)$. Denoting it by ∂_k, we obtain the chain complex $\{P_k(X), \partial_k\}$, viz., $\ldots \to P_k(X) \overset{\partial_k}{\to} P_{k-1}(X) \to \ldots$. Just as for any chain comples, we define its homology groups, i.e., $\text{Ker}\,\partial/\text{Im}\,\partial$, called the cellular homology of the complex. As it turns out, there exists the canonical isomorphism between the homology of the described chain complex and the singular homology of X. This is the basic assertion of the present section that helps reduce the computation of singular homology to the computation of the homology of a much simpler chain complex. The reduction turns out to be so effective that the majority of concrete homology computations are based on this very theorem. In particular, it immediately follows that cellular homology groups are homotopy-invariant and that singular homology groups of a finite complex are finitely generated.

4. Theorem on the Coincidence of Singular Homology with the Cellular Homology of a Finite Cell Complex.

THEOREM 3. *For a finite cell complex X, the singular homology groups $H_k(X)$ and the chain complex $\{P_k(X), \partial_k\}$, i.e., homology groups $\text{Ker}\,\partial_k/\text{Im}\,\partial_{k+1}$ (cellular homology groups) are isomorphic.*

First, we prove some auxiliary statements.

LEMMA 3. *For $k > 1$, the isomorphism $H_k(X^{k+1}, X^{k-2}) = H_k(X)$ holds.*

PROOF. Consider the triple of $(X^{k+1}, X^{k-2}, X^{k-3})$ and the corresponding exact sequence

$$H_k(X^{k-2}, X^{k-3}) \to H_k(X^{k+1}, X^{k-3}) \to H_k(X^{k+1}, X^{k-2}) \to H_{k-1}(X^{k-2}, X^{k-3}).$$

Its extreme terms are equal to zero, i.e., $H_k(X^{k+1}, X^{k-3}) = H_k(X^{k+1}, X^{k-2})$. Repeating the argument for the triple $(X^{k+1}, X^{k-3}, X^{k-4})$, we obtain $H_k(X^{k+1}, X^{k-3}) = H_k(X^{k+1}, X^{k-4})$. Getting the dimension increase further, we obtain the following chain of isomorphisms:

$$H_k(X^{k+1}, X^{k-2}) = H_k(X^{k+1}, X^{k-3}) = H_k(X^{k+1}, X^{k-4})$$
$$= \ldots H_k(X^{k+1}, X^0) = H_k(X^{k+1}) \quad \text{for} \quad k = 1.$$

If the skeleton X^0 consists of one point only, equality holds for $k = 1$. Any finite connected cell complex is homotopy-equivalent to a finite complex whose zero-dimensional skeleton consists of one point only. For verification, it suffices to consider all zero-dimensional cells of the original complex, and join each cell to one distinguished vertex (zero-dimensional cell $*$) by a continuous path; all paths should lie in the one-dimensional skeleton X^1. Next, we realize the homotopy contracting all zero-dimensional cells to one point (Fig. 13).

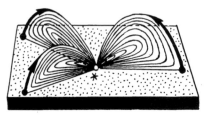

Figure 13

LEMMA 4. *There exists the isomorphism* $H_k(X) = H_k(X^{k+1})$.

PROOF. Let us prove that, for any $i < k+1$, the equality $H_i(X^{k+1}) = H_i(X^{k+2})$ holds. In fact, consider the exact sequence of the pair (X^{k+2}, X^{k+1}):

$$0 = H_{i+2}(X^{k+2}, X^{k+1}) \to H_i(X^{k+1}) \to H_i(X^{k+2}) \to H_i(X^{k+2}, X^{k+1}) = 0.$$

Hence, the required equality follows immediately; in particular, $H_k(X^{k+1}) = H_k(X^{k+2})$. Returning to the proof of Lemma 3, we obtain

$$H_k(X^{k+1}, X^{k-2}) = H_k(X^{k+1}) = H_k(X^{k+2}) = \ldots = H_k(X). \quad \text{Q.E.D.}$$

LEMMA 5. *The isomorphism* $\operatorname{Ker} \partial_k / \operatorname{Im} \partial_{k+1} = H_k(X^{k+1}, X^{k-2})$ *holds, where* ∂_k *are homomorphisms determining the complex of cellular chain groups.*

PROOF. Consider the commutative diagram

$$H_k(X^{k-1}, X^{k-2}) = 0$$
$$\downarrow$$
$$P_{k+1}(X) = H_{k+1}(X^{k+1}, X^k) \xrightarrow{\partial} H_k(X^k, X^{k-2}) \xrightarrow{i_*} H_k(X^{k+1}, X^{k-2}) \to$$
$$\downarrow \qquad \qquad \qquad \partial_{k+1} \qquad \qquad \downarrow \qquad \qquad \to H_k(X^{k+1}, X^k) = 0$$
$$\xrightarrow{\qquad\qquad} H_k(X^k, X^{k-1}) = P_k(X)$$
$$\downarrow \partial_*$$
$$H_{k-1}(X^{k-1}, X^{k-2}) = P_{k-1}(X).$$

Here, the line is a segment of the exact sequence of the triple (X^{k+1}, X^k, X^{k-2}), the column is a segment of the sequence of the triple (X^k, X^{k-1}, X^{k-2}), while the homomorphisms i_*, j_* are induced by the corresponding embeddings of the pairs i, j. It follows from diagram commutativity that $\partial_{k+1} = j_* \partial$. Recall that $P_\alpha(X) = H_\alpha(X^\alpha, X^{\alpha-1})$. Since the line and the column are segments of exact sequences, i_* is an epimorphism, whereas j_* is a monomorphism. Hence,

$$H_k(X^{k+1}, X^{k-2}) = H_k(X^k, X^{k-2})/\operatorname{Ker} i_* = H_k(X^k, X^{k-2})/\operatorname{Im} \partial.$$

Since j_* is a monomorphism,

$$H_k(X^k, X^{k-2})/\operatorname{Im} \partial = j_* H_k(X^k, X^{k-2})/j_* \operatorname{Im} \partial$$
$$= \operatorname{Im} j_* / \operatorname{Im} j_* \partial = \operatorname{Ker} \partial_k / \operatorname{Im} j_* \partial = \operatorname{Ker} \partial_k / \operatorname{Im} \partial_{k+1}.$$

Here, we have made use of the identities $\operatorname{Im} j_* = \operatorname{Ker} \partial_k$ (by exactness), $j_* \partial = \partial_{k+1}$ (meaning diagram commutativity). Thus,

$$H_k(X^{k+1}, X^{k-2}) = \operatorname{Ker} \partial_k / \operatorname{Im} \partial_{k+1}.$$

The lemma is proved.

The Proof of Theorem 3. From Lemmas 3 and 5, we obtain

$$H_k(X) = \operatorname{Ker} \partial_k / \operatorname{Im} \partial_{k+1} = H_k(X^{k+1}, X^{k-2}). \quad \text{Q.E.D.}$$

5. Geometric Definition of Cellular Homology Groups.

We need to clarify the geometric meaning of the operator ∂_k in the chain complex $\{P_k(X), \partial_k\}$. Consider two cells σ^{k-1} and σ^k in X. We assume that their orientations are fixed, and are compatible with the characteristic mappings $\chi : D^k \to X, \bar{\chi} : D^{k-1} \to X$. Consider the continuous mapping $\partial D^k = S^{k-1} \xrightarrow{\chi} X^{k-1}/X^{k-2}$, i.e., map the cell boundary onto the factor space of the $(k-1)$-dimensional skeleton relative to the $(k-2)$-dimensional skeleton. This factor space is homeomorphic to the union of $(k-1)$-dimensional spheres. Since, from the $(k-1)$-dimensional skeleton X^{k-1}, we have distinguished the cell σ^{k-1}, it is sent by the quotient map $K^{k-1} \xrightarrow{\alpha} X^{k-1}/X^{k-2}$ onto the sphere S^{k-1} from the union X^{k-1}/X^{k-2}. Thus, with one sphere distinguished from the union of spheres, we may define the natural projection of the whole union onto this sphere, i.e., the distinguished sphere remains fixed, while the others are mapped into the distinguished point (Fig. 14). The constructed mapping of the boundary of the ball D^k into the union of spheres and the subsequent projection onto the distinguished sphere determine a continuous mapping $S^{k-1} \to S^{k-1}$ and, in particular, establish a homomorphism of the group $\mathbf{Z} = H_{k-1}(S^{k-1})$ into itself. Each homomorphism $\mathbf{Z} \to \mathbf{Z}$ is uniquely determined by an integer m which is the image of the identity element of $\mathbf{Z}$. The integer is called the degree of a mapping. If the constructed mapping is smooth as in the examples below, then m coincides with the usual degree of a smooth mapping, defined for mappings of closed and orientable manifolds of the same dimension [1].

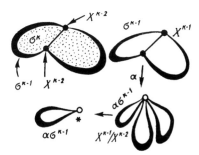

Figure 14

We have associated each pair of the cells σ and σ^{k-1} with a certain integer called the *incidence number* of cells, and generally denoted by $[\sigma^k:\sigma^{k-1}]$. The definition makes it clear that the integer depends on the chosen orientation of the cells; one of orientations being changed, there is change in sign of $[\sigma^k:\sigma^{k-1}]$.

THEOREM 4. *Let σ^k be an arbitrary generator of the group $P_k(X)=H_k(X^k, X^{k-1})$. Then the action of the boundary operator ∂ on σ^k is given by the formula $\partial\sigma^k=\Sigma[\sigma^k:\sigma^{k-1}]\sigma^{k-1}$, in which the sum is taken over all $(k-1)$-dimensional cells σ^{k-1} of the complex X.*

This statement yields a clear geometric interpretation of the boundary operator ∂ introduced above in the algebraic language for groups of cellular chains. If the cell σ^{k-1} and closure of the cell σ^k are disjoint, then $[\sigma^k:\sigma^{k-1}]=0$.

Proof of Theorem 4. Consider two triples $(D^k, S^{k-1}, \varnothing)$ and (X^k, X^{k-1}, X^{k-2}), and a continuous mapping $(D^k, S^{k-1}, \oplus)\to(X^k, X^{k-1}, X^{k-2})$, where $D^k\to X^k$ is the characteristic mapping of the cell σ^k, and $S^{k-1}\to X^{k-1}$ is the restriction of this mapping to the ball boundary. The exact sequences of these triples can be naturally organized into the commutative diagram

$$
\begin{array}{ccc}
0 & \mathbf{Z} & \mathbf{Z} \\
\| & \| & \| \\
\end{array}
$$

$$H_k(D^k)\to H_k(D^k, S^{k-1})\xrightarrow{j} H_{k-1}(S^{k-1})\to 0$$
$$\qquad i_*\downarrow \qquad\qquad\qquad\qquad \downarrow\phi_*$$
$$H_k(X^k, X^{k-1})\xrightarrow{\hat{c}} H_{k-1}(X^{k-1}, X^{k-2})$$
$$\qquad\| \qquad\qquad\qquad\qquad \|$$
$$P_k(X) \qquad\qquad\qquad P_{k-1}(X).$$

Consider the element $1\in\mathbf{Z}=H_k(D^k, S^{k-1})$. Under the homomorphism i_*, this generator is mapped onto the cellular chain $1\cdot\sigma^k\in P_k(X)$, and, after applying ∂, is carried into $1\cdot\partial\sigma^k$. Let us see how this generator shifts along the upper side of the square. Under the mapping j, the element 1 is sent into $1\in\mathbf{Z}=H_{k-1}(S^{k-1})$, since j is an isomorphism. The subsequent mapping φ_* carries the generator of $\mathbf{Z}$ into a certain element of $\mathbf{Z}\oplus\ldots\oplus\mathbf{Z}=H_{k-1}(X^{k-1}/X^{k-2})=H_{k-1}(\vee S^{k-1})$. The cell σ^{k-1} corresponds to each generator of the group $P_{k-1}(X)=H_{k-1}(\vee S^{k-1})$. It is clear that the coefficient of this cell in the image of the identity element under φ_* is equal to the degree of the composition mapping $S^{k-1}\to S^{k-1}$, i.e., to the coefficient $[\sigma^k:\sigma^{k-1}]$. The theorem is proved.

Thus, we have obtained a relatively simple rule for calculating the singular homology groups of a cell complex. It suffices to consider a chain complex of cellular chains uniquely determined by a cellular structure of X, and then to write out explicitly the boundary operators, for which it is sufficient to compute the incidence numbers of cell pairs of consecutive dimensions. Next, we should compute the homology groups of the obtained complex of groups.

This construction is so visual (and often calculated easily) that sometimes it is made the basis for determining cellular homology groups.'

DEFINITION 3. Let X be a finite cell complex, $P_k(X)$ cellular chain groups, and $\partial_k : P_k(X) \to P_{k-1}(X)$ the homomorphisms determined by the formula $\partial_k \sigma^k = \Sigma[\sigma^k : \sigma^{k-1}]\sigma^{k-1}$. Then the homology groups $\mathrm{Ker}\,\partial_k/\mathrm{Im}\,\partial_{k+1}$ of X are called *the cellular homology groups of the complex.*

To calculate these groups, one need not necessarily know the singular homology of X, since all the objects in Definition 2 admit a purely geometric description in terms of cells, characteristic mappings, and incidence coefficients. Theorem 3 is reformulated as follows: The singular and cellular homology groups of a finite cell complex are isomorphic. Corollary: If the same space X is represented as a cell complex in two ways, then the cellular homology groups of X do not depend on a cellular decomposition, since they are isomorphic to singular homology groups. Thus, to compute the homology of the space X, we should select its simplest representation in the form of a cell complex and calculate the cellular homology.

6. The Simplest Examples of the Computation of Cellular Homology Groups

EXAMPLE 1. The *sphere* S^n admits the simplest cellular decomposition $\sigma^0 \cup \sigma^n$, where σ^0 is a point and σ^n its complement in the sphere. It is clear that, for $n \geqslant 1$, there is $H_i(S^n) = \mathbf{Z}$ when $i = 0, n$, and $H_j(S^n) = 0$ when $j \neq 0, n$.

EXAMPLE 2. The *real projective space* RP^n. Recall that one of its realizations is the set of sequences of the form $x = (x_0, x_1, \ldots, x_n)$, where x_i are real numbers and at least one coordinate is non-zero, the sequence being considered up to a non-zero multiplier. The simplest cellular decomposition of RP^n is arranged as follows: As the cell σ^k, we should take all sequences x for which $x_k \neq 0$, $x_{k+1} = \ldots = x_n = 0$. Then, in each dimension k, we obtain exactly one cell σ^k, i.e., $RP^n = \sigma^0 \cup \sigma^1 \cup \ldots \cup \sigma^n$. Therefore, $P_k(RP^n) = \mathbf{Z}$. Now, it remains to compute the boundary operator $\partial_k : \mathbf{Z} \to \mathbf{Z}$. Fig. 15 shows the closure of the cell

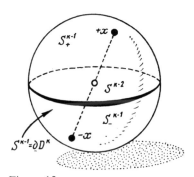
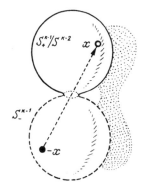

Figure 15

σ^k, i.e., $\mathbf{R}P^k$ is represented in the form of a k-dimensional ball whose boundary, the sphere S^{k-1}, has been factorized by the action of the group $\mathbf{Z}_2$, i.e., the generator of this group is represented by the transformation $x \to -x$, reflection of the sphere in the origin. In other words, $\mathbf{R}P^k$ is obtained from the ball D^k when diametrically opposite points are identified on its boundary, while the boundary of D^k is mapped to the quotient $\mathbf{R}P^{k-1}/\mathbf{R}P^{k-2} = S^{k-1}$ in the following way. Represent S^{k-1} as the union of three disjoint subsets S^{k-1}_+, S^{k-2}, S^{k-1}_-, where S^{k-1}_+ and S^{k-1}_- are open upper and lower hemispheres, and, respectively, S^{k-2} is the equator. The mapping $\partial D^k = S^{k-1} \to S^{k-1} = S^{k-1}_+/S^{k-2}$ is arranged as $x \to x$, if $x \in S^{k-1}_+$; $x \to *$ for $x \in S^{k-2}$, $-x \to x$ for $-x \in S^{k-1}_-$. Thus, we obtain the mapping $h: S^{k-1} \to S^{k-1} = S^{k-1}_+/S^{k-2}$, a diffeomorphism on each of the subsets S_+, S_-. It remains to find the *degree of the mapping*. It is clear that the inverse image of each point $x \in S_+$ is two points, viz., the point itself and its diametrically opposite on the sphere S^{k-1}. Hence, the unknown degree is either equal to two, or zero, depending on change in orientation of S^{k-1} under the mapping $x \to -x$. Therefore, we must find the degree of the auxiliary mapping $\alpha: S^{k-1} \to S^{k-1}$, where $\alpha(x) = -x$.

LEMMA 6. *The degree of the mapping* $\alpha(x) = -x$ *of the sphere* S^{k-1} *into itself is equal to* $(-1)^k$.

It follows immediately that the degree of a mapping h is equal to 2 when k is even, and zero when k is odd. Thus, for odd k, the coefficient $[\sigma^k : \sigma^{k-1}]$ is equal to zero, while, for even k, it is equal to 2. It means that the boundary operators in the chain complex $\{P_k(X), \partial_k\}$ are of the form $\partial \sigma^{2r} = 2\sigma^{2r-1}$, $\partial \sigma^{2r-1} = 0$. Thus, we have proved the following statement.

PROPOSITION 1. *The singular and cellular homology* RP^n *are of the form*

H_0	H_1	H_2	H_3	H_4	...	H_{n-1}	H_n	
$\mathbf{Z}$	$\mathbf{Z}_2$	0	$\mathbf{Z}_2$	0	...	$\mathbf{Z}_2$	0	for even n;
$\mathbf{Z}$	$\mathbf{Z}_2$	0	$\mathbf{Z}_2$	0	...	0	$\mathbf{Z}$	for odd n.

In conclusion, we prove Lemma 6. Consider the sphere S^{k-1}, and fix an arbitrary orthoframe $e(x) = (e_1, \ldots, e_{k-1})$ at a point x. Under the mapping $\alpha: x \to -x$, this frame is carried into $e(-x) = (-e_1, \ldots, -e_{k-1})$ (we consider the sphere to be standardly embedded into the Euclidean space). Compare orientations induced on the sphere by the two frames. Join the points x and $-x$ by such a meridian γ that, at x, the velocity vector is e_{k-1} (Fig. 16). Perform a smooth deformation (translation) of the frame $e(x)$ from x to $-x$, moving along the path γ, so that e_{k-1} is always tangent to γ. Then, at $-x$, we obtain two frames $-e_1, \ldots, -e_{k-2}, -e_{k-1}$ and $-e_1, \ldots, -e_{k-2}, +e_{k-1}$. It is obvious that their mutual orietation is determined by the sign $(-1)^{k-2}$. The lemma is proved.

EXAMPLE 3. A two-dimensional compact connected closed and orientable manifold M^2_g of genus g is homeomorphic to a two-dimensional sphere with g

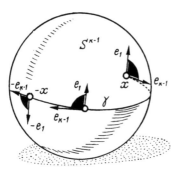

Figure 16

handles, and admits the cellular decomposition $\sigma^0 \cup \bigcup_{i=1}^{2g} \sigma_i^1 \cup \sigma^2$. It follows that the homology of M_g^2 is of the form $H_0 = \mathbf{Z}, H_1 = \mathbf{Z} \oplus \ldots \oplus \mathbf{Z}$ (to $2g$ addends), $H_2 = \mathbf{Z}$. Find the homology groups of a non-orientable *two-dimensional manifold*.

The above-determined homology groups do not exhaust "homology invariants" which sometimes admit distinguishing cell complexes. We can construct a homology theory, using chains with coefficients in an arbitrary Abelian group A. This means that the chains should be viewed as linear combinations with coefficients from A. It is clear that, in this case, all the constructions are carried out mechanically too, thus enabling us to determine the groups $H_k(X, A)$, referred to as homology groups with coefficients in the group A. The results for cellular homology hold for the groups A and $A = \mathbf{Z}$ considered above. In particular, these groups can be determined as chain complex homology constructed from the groups of cellular chains $\{\Sigma a_i \sigma_i^k, a_i \in A\} = P_k(X, A)$ with coefficients in A.

Homology groups $H_k(X)$ studied above are denoted by $H_k(X, \mathbf{Z})$ in new notation. For brevity, we will often omit the notation of a coefficient group provided the final result does not depend on the choice of A.

PROBLEM. For Examples 1–3 (see above), compute the homology groups with the coefficients in the groups $A = \mathbf{R}$, Q, $\mathbf{Z}_2$, $\mathbf{Z}_p$, where R is a field of real numbers, $\mathbf{Q}$ is a field of rational numbers, and p is a prime number.

2. Cohomology Groups and Obstructions to Extending Mappings

1. Singular Cochains and the Operator δ

Let X be a cell complex and $C_k(X)$ a group of k-dimensional singular chains of a space X. Let A be an Abelian group.

DEFINITION 1. A *singular cochain* of a space X with coefficients in the group A is referred to as a homomorphism of $C_k(X)$ into A. The natural operation of adding cochains transforms a set of cochains into an Abelian group, denoted by $C^k(X, A)$ and called a group of cochains of the space X. Consider the boundary operator $\partial : C_k(X) \to C_{k-1}(X)$. Let $h \in C^{k-1}(X, A)$ be an arbitrary cochain, i.e., a homomorphism $h : C_k(X) \to A$. Then, the cochain $\delta h \in C^k(X, A)$, given by the formula $\delta h(\alpha) = h(\partial \alpha)$, is determined uniquely, viz., $\delta h : C_k(X) \to A$. In other words, the cochain δh is determined by the diagram

$$\alpha \in C_k(X) \xrightarrow{\quad i \quad} C_{k-1}(X).$$

Sometimes, the operator $\delta : C^{k-1}(X, A) \to C^k(X, A)$ is denoted by δ_{k-1}.

DEFINITION 2. The operator $\delta_{k-1} : C^{k-1} \to C^k$ is called *coboundary*. It is conjugate to the boundary operator ∂.

Since $\partial^2 = 0$, $\delta^2 = 0$. Therefore, we obtain a sequence of groups and connecting homomorphisms of the form

$$C^0(X, A) \xrightarrow{\delta_0} C^1(X, A) \xrightarrow{\delta_1} C^2(X, A) \to \dots,$$

where $\delta_k \delta_{k-1} = 0$. This sequence is called a *cochain complex*. Following the procedure described in the preceding section, consider the groups $\operatorname{Ker} \delta$ and $\operatorname{Im} \delta$, and construct a group $H^k(X, A) = \operatorname{Ker} \delta_k / \operatorname{Im} \delta_{k-1}$.

DEFINITION 3. Groups $H^k(X, A)$ are referred to as *cohomology groups* of the space X with coefficients in the Abelian group A. The elements of the group $B^k = \operatorname{Im} \delta_{k-1}$ are called *coboundaries*, and those of $Z^k = \operatorname{Ker} \delta_k$, *cocycles*.

If the space X is path-connected, then $H^0(X, A) = A$. Relative cohomology groups are naturally determined, just as in the homology case. Let Y be a closed subcomplex in a complex X, then $C_k(Y) \subset C_k(X)$. Let $C^k(X, Y)$ be a group of all homomorphisms $\alpha : C_k(X) \to A$ equal to zero on $C_k(Y)$. It is evident that $\delta C^k(X, Y) \subset C^{k+1}(X, Y)$, hence, there arise groups $H^k(X, Y, A) = \operatorname{Ker} \delta / \operatorname{Im} \delta$ referred to as *relative cohomology groups*. Just as in the homology case, there arises naturally the exact sequence of the pair and triple. Omitting the details of the construction, we give only the exact sequence of the pair (verify exactness!)

$$\dots \to H^k(X, Y, A) \to H^k(X, A) \to H^k(Y, A) \xrightarrow{\delta} H^{k+1}(X, Y, A) \to \dots$$

and the triple (X, Y, Z):

$$\dots \to H^k(X, Y, A) \to H^k(X, Z, A) \to H^k(Y, Z, A) \xrightarrow{\delta} H^{k+1}(X, Y, A) \to \dots.$$

Singular cohomology groups are homotopy invariant. Following the procedure described in Sec. 1, define cellular cohomology groups. To this end, we introduce groups of cellular cochains $P^k(X, A)$ defined as $H^k(X^k, X^{k-1}, A)$, where X^k is a k-dimensional skeleton of the cell complex X. The exact sequence of the triple (X^{k+1}, X^k, X^{k-1}) determines the coboundary operator $\delta: P^k(X, A) \to P^{k+1}(X, A)$. Cellular cohomology groups are defined as the groups $\text{Ker } \delta / \text{Im } \delta$ for the cochain complex $\{P^k(X, A), \delta\}$.

THEOREM 1. *For a finite cell complex X the singular cohomology groups and cellular cohomology groups are isomorphic, and they are finitely generated Abelian groups if such is the coefficient group.*

Proof is obtained just as in Section 1.

2. The Problem of the Extension of a Continuous Mapping from a Subspace to the Whole Space

To study variational problems in topology in the chapters to come, we have to solve the following problem.

Let Y be a closed subspace of a topological space X, and let be given a continuous mapping of this subspace into some space Z. The question arises as to the conditions under which this mapping can be extended to a continuous one of the whole space X into Z. It is obvious that such a continuous mapping does not always exist. There are topological obstructions which, in some cases, hinder extending a mapping from a subspace to the whole space. The study of all different versions is dealt with by obstruction theory. Here, we only give the results used in the sequel. We have already all the necessary material elaborated in the previous sections to give the required information.

3. The Obstruction to Extending Mappings.

Consider a finite cell complex K and a topological space X. Given a continuous mapping g of the $(n-1)$-dimensional skeleton K^{n-1} of the cell complex K (i.e., of the union of all its cells of dimension not exceeding $n-1$) into the space X. We want to extend it to a mapping f of the next skeleton K^n into X. For simplicity, we assume X to be either 1-connected, or, in case X is one-dimensional, to have a commutative fundamental group. Consider all n-cells σ^n composing the n-dimensional skeleton of a complex K. By its finiteness, there is a finite number of these cells. To construct an unknown extension, one must know how to extend the original mapping to each n-cell taken separately. Fix any cell σ^n. Its boundary has already been mapped by the mapping g into the space X. This mapping should be extended to the whole cell. Assume that the mapping under construction $f: K^n \to X$ coincides with the mapping g on the skeleton $K^{n-1} \subset K^n$. Let $\chi: D^n \to K$ be a characteristic mapping of σ^n, where D^n is a n-dimensional disk. There arises the composition mapping $S^{n-1} = \partial D^n \xrightarrow{\chi} K^{n-1} \xrightarrow{g} X$. Here, we make use of the fact that the

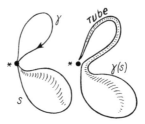

Figure 17

boundary of the cell σ^n is contained in the skeleton K^{n-1} according to the cell complex definition. The obtained mapping $g\chi : S^{n-1} \to X$ determines some element $[g\chi]$ of the homotopy group $\pi_{n-1}(X)$. This definition is correct due to restrictions placed on the space X. They guarantee that defining the elements of $\pi_{n-1}(X)$ does not depend on the choice of the basic (distinguished) point. In the general case, it is essential for the space X to be $(n-1)$-simple (see the definition in [4]), i.e., the fundamental group must act trivially (from the homotopy standpoint) on $\pi_{n-1}(X)$. See this action in Fig. 17. It is clear that in the above assumptions X is $(n-1)$-simple.

The mapping $g\chi : S^{n-1} \to X$ can be extended to that of the whole disk D^n if and only if $[g\chi]$ from $\pi_{n-1}(X)$ is zero. In the general case, each cell σ^n is associated with some element $[g\chi]$ of the Abelian group $G = \pi_{n-1}(X)$. Consider a group $P_n(K)$ of cellular chains of the complex K. It is apparent that the constructed correspondence $\sigma^n \to [g\chi] \in G$ can be naturally extended to a certain homomorphism of the chain group $P_n(K)$ into G. To this end, if suffices to construct a mapping for each cell and extend it to all chains by linearity. Thus, we determine a certain c_g^n from a group of cellular cochains.

DEFINITION 4. The cochain $c_g^n \in P^n(K)$ with the coefficients from the Abelian group $G = \pi_{n-1}(X)$ is called an *obstruction* to the extension of *the mapping g* from the skeleton K^{n-1} to the skeleton K^n.

LEMMA 1. *The mapping $g : K^{n-1} \to X$ can be extended to the continuous mapping $f : K^n \to X$ if and only if the cochain c_g^n is identically equal to zero.*

Proof follows from the definition of the obstruction.

LEMMA 2. *Let a space X be $(n-1)$-simple. The cochain $c_g^n \in P^n(K, \pi_{n-1}(X))$ is a cocycle, i.e., $\delta c_g^n = 0$. Therefore, the cochain c_g^n determines some cohomology class $C_g^n = [c_g^n]$ from the group $H^n(K, \pi_{n-1}(X))$.*

Since the basic ideas of the proof will be of no use in the sequel, the proof is omitted (see [4]). In applications we are going to encounter, the equality $\delta c_g^n = 0$ will follow from geometry.

THEOREM 2. *Let a space X be $(n-1)$-simple. The cocycle C_g^n is zero (as an element of a cohomology group) if and only if the original mapping $g : K^{n-1} \to X$*

$g:K^{n-1}\rightarrow X$ can be extended to the continuous mapping $f:K^n\rightarrow X$, the mapping g having been changed on the skeleton K^{n-1}, and kept intact on K^{n-2}

Unlike Lemma 2, the scheme of proof will be used in the applications, therefore, we give the proof.

PROOF. To construct a mapping f, we need a new notion closely connected with the obstruction c_g^n, viz., the notion of a difference cochain. Given two mappings $f,g:K^{n-1}\rightarrow X$ coinciding on the skeleton K^{n-2}, i.e., $f(x)=g(x)$ for $x\in K^{n-2}$. Let $\sigma^{n-1}\subset K^{n-1}$ be an arbitrary cell, and $\chi:D^{n-1}\rightarrow K$ its characteristic mapping. Since the boundary S^{n-2} of the disk D^{n-1} is mapped into the skeleton K^{n-2}, two composition mappings $f\chi$ and $g\chi$ map the sphere S^{n-2} into the complex K in the same way. The cell σ^{n-1} is mapped, generally speaking, in different ways, though these two images have common boundary, since they are glued together with respect to the image of the sphere S^{n-2} in K. Therefore, in the space X, we obtain a spheroid defining some element of the group $\pi_{n-1}(X)$ and measuring the deviation of the mapping f from the mapping g on σ^{n-1}. Thereby, we have associated each cell σ^{n-1} with an element of the group $\pi_{n-1}(X)$ (here, $(n-1)$-simplicity is used). We obtain a homomorphism of a group of the chains $P_{n-1}(K)$ in the group $\pi_{n-1}(X)$. i.e., $(n-1)$-dimensional cochain. This cochain is denoted by $d_{f,g}^{n-1}$, and it is called a *difference cochain* of two mappings f and g (Fig. 18). From the definition it follows that $d_{f,g}^{n-1}=0$ if and only if there exists a homotopy connecting f and g, and constant on the skeleton K^{n-2}, where f and g coincide.

LEMMA 3. *For any mapping $f:K^{n-1}\rightarrow X$ and any cochain $d\in P^{n-1}(K,\pi_{n-1}(X))$, one can always select such a mapping $g:K^{n-1}\rightarrow X$ which coincides with f on the skeleton K^{n-2}, and a cochain d is the difference cochain for f and g, i.e., $d=d_{f,g}^{n-1}$.*

PROOF. Consider an arbitrary cell σ^{n-1} from K and its image under the mapping f into X. Distinguish a sufficiently small ball in the center of the cell and, on considering its image under f in X, cut this image out (using the cellular approximation theorem [4]) from the image of the cell σ^{n-1}. Next, glue the spheroid, which is a representative of the element from the group $\pi_{n-1}(X)$, to the obtained hole. (Spheroid is a representative, the hole is not). The element is the value of a cochain d on the cell σ^{n-1} (Fig. 19). As a new

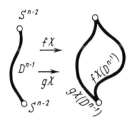

Figure 18

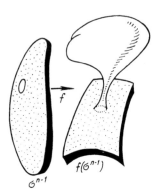

Figure 19

mapping g, take the mapping coincident with f everywhere except the distinguished ball, while on the ball it coincides with the mapping realizing the above-mentioned spheroid. Roughly speaking, the distinguished small ball bulges into a spheroid realizing the value of the cochain d on this cell. Performing this operation on each cell, we obtain some mapping g. Comparing g with the original mapping, we obtain, obviously, that $d = d_{f,g}^{n-1}$. The lemma is proved.

LEMMA 4. *The equality $\delta d_{f,g}^{n-1} = c_f^n - c_g^n$ holds.*

PROOF. By the definition of a coboundary operator in terms of cellular cochains and chains, there exists an equality

$$(\delta d_{f,g}^{n-1})\sigma^n = d_{f,g}^{n-1}(\delta\sigma^n) = \Sigma[\sigma^n : \sigma_i^{n-1}]d_i^{4n-1}][\sigma^n : \sigma_i^{n-1}]d_{f,g}^{n-1}(\sigma_i^{n-1}).$$

To calculate the incidence numbers, we must consider the following composition mapping

$$S^{n-1} = \partial D^n \xrightarrow{\chi} K^{n-1} \xrightarrow{\pi} K^{n-1}/K^{n-2} = \vee S_i^{n-1} \xrightarrow{p_i} S_i^{n-1},$$

where π is the natural factorization, p_i is the projection of the wedge of spheres onto the sphere indexed i. The degree of the obtained mapping $S^{n-1} \to S_i^{n-1}$ is called the *incidence number* $[\sigma^n : \sigma_i^{n-1}]$. It is easy to prove that the mapping $S^{n-1} \to K^{n-1}$ is homotopic to such a mapping, under which almost the whole sphere S^{n-1} except a finite number of small balls on this sphere is mapped into the skeleton K^{n-2}, while the small balls are mapped into the skeleton K^{n-1}, each small ball being mapped onto its cell σ_i^{n-1} with degree ± 1. The algebraic number of small balls mapped onto the cell σ_i^{n-1} (i.e., the sum of degrees ± 1) is exactly equal to the required incidence number (Fig. 20). This figure shows thick segments on the boundary of the ball, which denote small balls of dimension $n-1$, *mapped onto the cells σ_i^{n-1}*. Compute the value of the cochain $c_f^n - c_g^n$ on σ^n (Fig. 20), i.e., where the boundary cells σ_i^{n-1} will be mapped onto. At last, we must find out where the small balls distinguished on S^{n-1} will map under a composition mapping by means of f and g. Since the mappings f and g

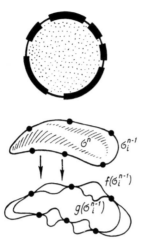

Figure 20

coincide on the skeleton K^{n-2}, to compute the value of the cochain on the cell σ^n, one must add up along the boundary $\partial \sigma^n$, i.e., on all the cells σ_i^{n-1}, the following spheroids from $\pi_{n-1}(X)$. On each cell σ_i^{n-1}, a spheroid must be taken, which coincides with the value of the difference cochain $d_{f,g}^{n-1}$ on the cell σ_i^{n-1}, the spheroid being taken as many times as the small balls mapped onto σ_i^{n-1}. It is obvious that this sum yields the incidence number, and the whole cell σ^n will, therefore, be associated with the sum $\Sigma[\sigma^n : \sigma_i^{n-1}] d_{f,g}^{n-1}(\sigma_i^{n-1})$ coinciding with $(\delta d_{f,g}^{n-1}) \sigma^n$. The proof is complete.

Revisit the proof of Theorem 2. Let the mapping $g : K^{n-1} \to X$ extend to the mapping $f : K^n \to X$ without any change on the skeleton K^{n-2} but, possibly, with a change on the skeleton K^{n-1}. By Lemma 4, we obtain $\delta d_{f,g}^{n-1} = c_f^n - c_g^n$, Since f has been defined on K^n, $c_f^n = 0$, therefore, $c_g^n = -\delta d_{f,g}^{n-1}$, and we have $C_g^n = 0$. Prove the converse statement. Given $C_g^n = 0$, i.e., $c_g^n = \delta d$, where d is a cochain. According to Lemma 3, there exists a mapping $f : K^{n-1} \to X$, coinciding with the mapping g on the skeleton K^{n-2}, such that $-d = d_{f,g}^{n-1}$. Then, by Lemma 4, we obtain that $c_f^n = c_g^n + \delta d_{f,g}^{n-1} = c_g^n - \delta d = 0$. Therefore, the mapping f extends to the mapping $K^n \to X$. Thus, we have extended the mapping g, having adjusted it, probably, on the skeleton K^{n-1}, but not on K^{n-2}. Q.E.D.

4. Cases of the Retraction of a Space on a Subspace Homeomorphic to a Sphere

Let us prove the retraction theorem which will be of use in the study of variational problems. Consider an Abelian group (an additive group)

$U = \mathbf{R}^1(\text{mod } 1)$, i.e., a circle, as a group of coefficients of cellular homology theory.

THEOREM 3 (due to Hopf). *Given an embedding i of the sphere S^{n-1} into a finite n-dimensional cell complex K. Let a homomorphism $i_*: H_{n-1}(S^{n-1}, U) \to H_{n-1}(X, U)$ induced by this embedding be a monomorphism. Then the sphere $S_0^{n-1} = iS^{n-1}$ is a* **retract** *of K, i.e., there exists a continuous mapping $f: K \to S_0^{n-1}$ identical on the sphere S_0^{n-1}.*

A constructive version of the proof, indispensable for an explicit construction of concrete *retractions*, is due to T. N. Fomenko (see [67]). We must construct the inverse continuous mapping $f: K \to S^{n-1}$, so that its restriction on the sphere iS^{n-1}, embedded in K, is the identity mapping of the sphere iS^{n-1} into S^{n-1}. To construct such a mapping, we apply the obstruction theory. First of all, for the sphere S^{n-1} choose the simplest cellular decomposition into a sum of two cells: a zero-dimensional cell $*$ and a $(n-1)$-dimensional cell σ^{n-1}. Therefore, we can construct an unknown mapping so that it is cellular and it contracts the whole skeleton K^{n-2} to a point on the sphere S^{n-1}. In other words, it suffices to construct a continuous mapping of a quotient complex K/K^{n-2} into the sphere S^{n-1}; here, $K = K^n$. To apply the obstruction theory, $(n-1)$-simple space must be considered as X. Take $X = S^{n-1}$, then the condition of $(n-1)$-simplicity is, obviously, satisfied. In fact, if $n-1 > 1$, then $\pi_1(X) = 0$; but if $n-1 = 1$, then the group $\pi_1(S^1) = \mathbf{Z}$ is Abelian, and acts on itself identically. As the skeleton K^{n-2} has been contracted to a point, we can assume that the skeleton K^{n-1} is coincident with the wedge of $(N-1)$-dimensional spheres $S_0^{n-1} \vee S_1^{n-1} \vee \ldots \vee S_k^{n-1}$. It is clear that the sphere iS^{n-1} may be regarded as one of them. For definiteness, let it be the sphere S_0^{n-1}. Represent each of the spheres S_i^{n-1} in the form of simplest cellular decomposition, and denote the corresponding $(n-1)$-dimensional cells by σ_i^{n-1}. We must construct such a continuous mapping $f: K \to S^{n-1}$ that its restriction on S_0^{n-1} is the identity mapping on S^{n-1}. Let $\sigma_1^n, \ldots, \sigma_q^n$ be n-cells of the complex K. Then we can assume that $K = \sigma^0 \cup \sigma_0^{n-1} \cup \ldots \cup \sigma_k^{n-1} \cup \sigma_g^n$ (Fig. 21).

Prior to the above-mentioned procedure, we have to construct a certain mapping of the skeleton K^{n-1} into S^{n-1}. Now, we construct some mapping g_1 which, generally speaking, will not extend to the mapping of the whole K into S^{n-1}, though it can be used to construct another, now final, mapping g_2, extending to the mapping $f: K \to S^{n-1}$. Since the skeleton K^{n-1} is the wedge of spheres, it suffices to prescribe the mapping g_1 on each of them. On the distinguished sphere S_0^{n-1}, we prescribe the identity mapping $S_0^{n-1} \to S^{n-1}$, while other spheres map onto a distinguished point $* \in S^{n-1}$. This mapping is continuous, but it is easy to give examples showing that this mapping must not necessarily extend to a continuous mapping of the whole complex into the sphere. Calculate the obstruction $c_{g_1}^n$ to the mapping g_1.

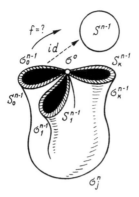

Figure 21

Consider an arbitrary n-cell σ^n. To compute $c_{g_1}^n(\sigma^n)$, we must consider the composition mapping of the $(n-1)$-dimensional sphere, cell boundary, into the sphere S^{n-1}, and find its degree. Here, we make use of $\pi_{n-1}(S^{n-1}) = \mathbf{Z}$ and homotopy classes of the mappings of a sphere into a sphere being classified by the degree of a mapping. This composition mapping is the composition of a characteristic mapping and the mapping $g_1 : K^{n-1} \to S^{n-1}$. The part of the boundary sphere $\partial \sigma^n$, which has been mapped into the wedge $S_1^{n-1} \vee \ldots \vee S_k^{n-1}$, i.e., non-incident with the distinguished sphere S_0^{n-1}, will, finally, map onto a distinguished point $*$ on the sphere S^{n-1}. Therefore, these parts of the sphere $\partial \sigma^n$ do not contribute to forming the unknown degree (the contribution is zero). The rest of the boundary sphere $\partial \sigma^n$ will map into the distinguished sphere S_0^{n-1}, "winding up" on it as many times as the value of the incidence number $[\sigma^n : \sigma_0^{n-1}]$ is and mapping into the sphere S^{n-1} by means of the identity mapping. As a result, the boundary sphere $\partial \sigma^n$ maps into the sphere S^{n-1} with a degree equal to the incidence number $[\sigma^n : \sigma_0^{n-1}]$. Thus, we have computed that $c_{g_1}^n(\sigma^n) = [\sigma^n : \sigma_0^{n-1}]$. We state that this cochain is representable in the form of a coboundary from some $(n-1)$-dimensional cochain d.

LEMMA 5. *In the assumptions of Theorem 3, the cochain $c_{g_1}^n$ is a coboundary, i.e.,* $c_{g_1}^n = \delta d$, *where* $d \in P^{n-1}(K, \pi_{n-1}(S^{n-1}))$ *and* $d(\lambda \sigma_0^{n-1}) = 0$, $0 \leqslant \lambda \leqslant 1$.

For simplicity, consider the case where among n-cells of the complex K there is only one cell σ^n having a non-zero incidence number with the distinguished cell σ_0^{n-1}. Roughly, our further reasoning will develop in the following way. We aim at constructing a $(n-1)$-dimensional cochain d on the wedge of spheres, this cochain being equal to zero on the distinguished cell σ_0^{n-1} (more precisely, on the elementary chain $1 \cdot \sigma_0^{n-1}$) and, generally speaking, non-trivial on the remaining spheres of the wedge, i.e., we want to "take off" the cochain from the distinguished sphere S_0^{n-1}. Take advantage of the fact that the cycle S_0^{n-1} is not homologous to zero in the complex K for the group of coefficients U. Recall that the sphere embedding-induced homomorphism has no kernel. Compute the boundary on the cell σ^n. We have $\partial\sigma^n = [\sigma^n:\sigma_0^{n-1}]\sigma_0^{n-1} + \ldots + [\sigma^n:\sigma_k^{n-1}]\sigma_k^{n-1}$. Denoting $[\sigma^n:\sigma_i^{n-1}]$ by a_i, we can write down $\partial\sigma^n = a_0\sigma_0^{n-1} + a_1\sigma_1^{n-1} + \ldots + a_k\sigma_k^{n-1}$. Thus, the chain $-a_0\sigma_0^{n-1}$ is homologous to the chain $a_1\sigma_1^{n-1} + \ldots + a_k\sigma_k^{n-1}$. Therefore, the cycles defined by them coincide as elements of the homology group $H_{n-1}(K,\mathbf{Z})$.

It follows from the condition of the theorem that the cycle $\lambda\sigma_0^{n-1}$, where $0 < \lambda < 1$, is non-zero in the group $H_{n-1}(X,U)$. Consider three following cases: (a) All integers $a_1, \ldots, a_k$ are coprime numbers, (b) integers $a_1, \ldots, a_k$ have the greatest common divisor p different from unity, a_0/p not being an integer, i.e., p does not divide a_0, (c) integers $a_1, \ldots, a_k$ have the greatest common divisor p different from zero, and p divides a_0.

Consider (a).

Since $a_1, \ldots, a_k$ are coprime, there exist such integers $x_1, \ldots, x_k$ that $a_0 = x_1 a_1 + \ldots + x_k a_k$. Define the $(n-1)$-dimensional cochain d on the skeleton K^{n-1} in the following way: $d(\lambda\sigma_i^{n-1}) = \lambda x_i$, $x_i \in \mathbf{Z} = \pi_{n-1}(S^{n-1})$, $1 \leq i \leq k$ and $d(\lambda\sigma_0^{n-1}) = 0$. Here, the integers $x_1, \ldots, x_k$ are realized as the degrees of mappings of the spheres S_i^{n-1} into the sphere S_{n-1}. It is clear that $\delta d = c_{g_1}^n$. In fact,

$$(\delta d)\sigma^n = d(\delta\sigma^n) = \sum_{i=1}^{k} a_i d\sigma_i^{n-1} = \sum_{i=1}^{k} x_i a_i = a_0 = c_{g_1}^n(\sigma^n).$$

Thus, in case (a), the lemma is proved.

In case (b), consider an equality $\partial(\lambda\sigma^n) - \lambda a_0\sigma_0^{n-1} = \lambda p(b_1\sigma_1^{n-1} + \ldots + b_k\sigma_k^{n-1})$, where $a_i = pb_i$, $1 \leq i \leq k$. Here, $b_1, \ldots, b_k$ are coprime numbers. As a_0/p is not an integer, then, using a coefficient group U, we obtain that there exists a zero-cycle $\dfrac{a_0}{p}\sigma_0^{n-1}$ (under $\lambda = 1/p$) in the homology $H_{n-1}(S^{n-1},U)$, which becomes homologous to zero in $H_{n-1}(K,U)$, since $\dfrac{a_0}{p}\sigma_0^{n-1} \equiv \partial\left(\dfrac{1}{p}\sigma^n\right)$ (mod 1). It contradicts the monomorphism of $H_{n-1}(S^{n-1},U) \to H_{n-1}(K,U)$. Therefore, (b) is not actually realized.

Finally, consider (c). Since p is the greatest common divisor of $a_1, \ldots a_k$, $a_i = pb_i$, $1 \leq i \leq k$, $a_0 = mp$, $m \in \mathbf{Z}$, and $b_1, \ldots, b_k$ are coprime numbers. Hence, there exist integers $x_1, \ldots, x_k$ such that $m = x_1 b_1 + \ldots + x_k b_k$. Realize x_i as the degrees of mappings of the spheres S_i^{n-1} into the sphere S^{n-1} and define a cochain $d: d(\sigma_i^{n-1}) = x_i \in \pi_{n-1}(S^{n-1})$, $d(\sigma_0^{n-1}) = 0$. It is stated that $\delta d = c_{g_1}^n$. In fact, $(\delta d)\sigma^n = d(\partial\sigma^n) = d(a_0\sigma_0^{n-1} + \ldots + a_k\sigma_k^{n-1}) = a_1 x_1 + \ldots + a_k x_k = mp = a_0 = c_{g_1}^n\sigma^n)$. The lemma is proved.

Revisit the proof of Theorem 3. By Lemma 3 for the mappings $g_1: K^{n-1} \to S^{n-1}$ and the cochain d constructed above, i.e., such that $\delta d = c_{g_1}^n$, we can choose such a continuous mapping $g_2: K^{n-1} \to S^{n-1}$ that the equality $d = d_{g_1,g_2}^{n-1}$ is satisfied. By Lemma 4, we have $\delta d_{g_1,g_2}^{n-1} = \delta d = c_{g_1}^n - c_{g_2}^n$, i.e., $c_{g_2}^n = 0$.

Thus, we have obtained the mapping g_2 of a $(n-1)$-dimensional skeleton K into S^{n-1} which extends to a continuous mapping $f: K^n \to S^{n-1}$. It remains to verify that the mapping f is identity on the sphere S_0^{n-1}. It follows from the proof of Lemma 3. In fact, g_2 is constructed in the following way. It differs from g_1 only in $(n-1)$-dimensional cells on which the cochain is non-zero. But the cochain d has been specially constructed so that it is zero on the chains of the form $\lambda \sigma_0^{n-1}$; hence, g_2 and g_1 are identity mappings of the sphere S_0^{n-1} into the sphere S^{n-1}. If there are several n-cells incident with the cell σ_0^{n-1}, the proof is performed according to the same scheme, and we leave it to the reader. The theorem is proved.

The given proof is awkward in spite of its simple geometry. We have chosen this approach since it will be of use in the sequel, when we have to construct a concrete retraction of some complex on a sphere. An explicit construction of the required retraction becomes elementary after the above reasoning, though it seems appropriate to give another proof of Theorem 3 here, which, being shorter and perhaps not so obvious, yet clarifies general topological reasons for the required retraction. It is true that we have to resort to the properties of the so-called *Eilenberg-Maclane complexes*. Fix an integer $n \geqslant 1$ and a group π assumed to be Abelian for $n > 1$. Then there exists such a cell complex denoted by $K(\pi, n)$ and referred to as the Eilenberg-Maclane space that all its homotopy groups are trivial except $\pi_n(K(\pi, n))$ isomorphic to the group π. An important statement is valid: The set $\Pi(K, K(\pi, n))$ of homotopy classes of mappings of the finite cell complex K into the space $K(\pi, n)$ is in a one-to-one correspondence with the set of elements of the cohomology group $H^n(K, \pi)$. See the proof in [4], for example. Theorem 3 is deduced in the following way. Consider the sphere S^{n-1}, and embed it in the space $K(\mathbf{Z}, n-1)$ in the form of the $(n-1)$-dimensional skeleton. To accomplish this, it is necessary to glue some cells to the sphere to "eliminate" its all homotopy groups, beginning with the dimension n. Mean while, we make use of the fact that $\pi_{n-1}(S^{n-1}) = \mathbf{Z}$. The first homotopy group to be annihilated is $\pi_n(S^{n-1})$. Hence, the cells of dimensions $n+1$, $n+2$, etc., must be glued to the sphere S^{n-1}. Removing all higher homotopy groups consecutively, we obtain some complex B for which $\pi_i(B) = 0$ when $i \neq n-1$ and $\pi_{n-1}(B) = \mathbf{Z}$. It is well known that any two polyhedra which are the spaces of type $K(\pi, n-1)$ are homotopy equivalent for the given $n-1$ and π. It is clear that $H^{n-1}(K(\mathbf{Z}, n-1), \mathbf{Z}) = \mathbf{Z}$, $K(\mathbf{Z}, n-1) = \sigma^0 \cup \sigma^{n-1} \cup \sigma^{n+1} \cup \sigma^{n+2} \cup \ldots$. Let $\alpha \in H^{n-1}(K(\mathbf{Z}, n-1), \mathbf{Z})$ be the generator of $\mathbf{Z}$. Then (see [4]) the function carrying the cohomology class $f^* \alpha \in H^{n-1}(K, \mathbf{Z})$ to the mapping $f: K \to K(\mathbf{Z}, n-1)$ defines a one-to-one correspondence between $H^{n-1}(K, \mathbf{Z})$ and the set $\Pi(K, K(\pi, n))$ of homotopy classes of the mappings of K into $K(\pi, n)$ for any cell complex K. The condition of Theorem 3 by which the homomorphism $i_*: H_{n-1}(S^{n-1}, U) \to H_{n-1}(K, U)$ has no kernel can be reformulated equivalently in the cohomology language, viz., the homomorphism $i^*: H^{n-1}(K, \mathbf{Z}) \to H^{n-1}(S^{n-1}, \mathbf{Z})$ induced by the embedding is an epimorphism. Therefore, there exists such an element $\beta \in H^{n-1}(K, \mathbf{Z})$ that $i^*\beta = \alpha \in H^{n-1}(S^{n-1}, \mathbf{Z})$. By virtue of the foregoing, there is such a continuous mapping $f: K \to K(\mathbf{Z}, n-1)$ that $f^*\alpha = \beta$. Since $(fi)^*\alpha = \alpha$, the restriction of the mapping f on the sphere iS^{n-1} embedded in K is homotopy equivalent to the mapping of the sphere S^{n-1} into itself. It remains to prove that the mapping f can be deformed into such a mapping that will remain identity on the sphere S^{n-1} and will map the whole complex K into it. As the complex K is, by assumption, n-dimensional and the complex $K(\mathbf{Z}, n-1)$ does not contain cells of dimension n, we can consider, according to the cellular approximation theorem [4], that the mapping f carries K into the $(n-1)$-dimensional skeleton of the space $K(\mathbf{Z}, n-1)$, i.e., into the sphere S^{n-1}. This completes the proof of Theorem 3.

3. Fibrations

1. Defining a Locally Trivial Fibration.

After reformulation in the language of geometry, many applied problems give rise to spaces E having a special structure with the main property formulated roughly as follows: *A space E is decomposed into a continuous union of its subspaces* (fibers) *homeomorphic to one and the same space F*; locally, this decomposition is arranged rather simply, viz., any subset of points in E, composed of "neighboring fibers" is homeomorphic to the direct product of the fiber F and some space (Fig. 22). Such fibration of E into the fibers F turns out to be extremely useful when we study the geometry of the space E. Now, we give a mathematical definition.

DEFINITION 1. Given topological spaces E, B, F, and a continuous mapping p of the space E into B. We say that the quadruple E, B, F, p is a *locally trivial fibration* if, for each point $b \in B$, there exist such an open neighborhood $U \subset B$ and homeomorphism $\varphi: p^{-1}U \to U \times F$, that the following diagram $p^{-1}U \longrightarrow U \times F$, where $\pi: U \times F \to F$ is the natural projection of the

$$\begin{array}{c} p^{-1}U \xrightarrow{\ \varphi\ } U \times F \\ {}_{p}\searrow \quad \swarrow_{\pi} \\ U \end{array}$$

direct product onto F, is commutative, i.e., $\pi\varphi = p$. By $p^{-1}U$, we denote the full

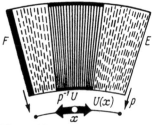

Figure 22

inverse image U under the mapping p, E being called the total space, F fiber, B base, p projection.

Sometimes, the mapping p itself is called a fiber bundle, a notation $p: E \overset{F}{\to}$

DEFINITION 2. The fibration $p: E \overset{F}{\to} B$ is said to be *trivial* if E is homeomorphic to $B \times F$, the diagram $E \overset{\alpha}{\longrightarrow} B \times F$ being commutative, i.e., $p = \pi\alpha$,

$$E \overset{\alpha}{\longrightarrow} B \times F$$
$$\underset{p}{\searrow} \quad \underset{\pi}{\swarrow}$$
$$B$$

where α is a homeomorphism, π projection. The latter means that, under the homeomorphism α, each fiber corresponds to one point of the base space (after projection). The origin of the term "local triviality" is seen from Definition 2: The fibration $p^{-1}U \overset{F}{\to} U$ is trivial for a sufficiently small neighborhood U.

2. Examples of Fibrations

EXAMPLE 1. **Direct products** $E = B \times F$ are fibrations relative to both projections $B \times F \to B$ and $B \times F \to F$ (Fig. 23).

EXAMPLE 2. A **Möbius band** E, where F is a segment I, B is a circle S^1, $p: E \to B$ *is a projection of a Möbius band* onto its median (Fig. 24). Since E is a non-orientable manifold, this fibration is not trivial (prove!).

EXAMPLE 3. Let $f: M^k \to N^p$ be a smooth mapping of a connected compact smooth manifold M onto a smooth compact manifold N, $k \geq p$, and f being a *regular mapping* in the sense that its differential df has maximal rank equal to p at each point of $x \in M$. Then, as is generally known (see, e.g., [1]), each fiber $F_y = f^{-1}(y)$, $y \in N$ is a smooth $(k-p)$-dimensional submanifold in M and, therefore, taking into consideration the implicit-function theorem, we obtain that the mapping f defines a locally trivial fibration.

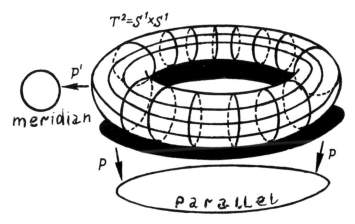

Figure 23

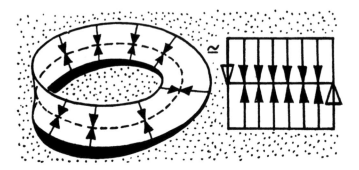

Figure 24

This example is particularly important, since most fibrations we are going to encounter are precisely of this type. Among them, a remarkable class related to factor spaces is singled out. Let G be a compact Lie group, and $H \subset G$ a closed subgroup [1]. Consider the set G/H of cosets gH with respect to the subgroup H. This set is naturally endowed with topology (close cosets are determined as multiplications of the subgroup H by close elements g, $g' \in G$). We obtain a mapping $p:G \to G/H$, where $p(g)=(gH)\in G/H$. The space G/H is naturally equipped with the structure of a smooth manifold [1]. The proof is omitted, since this fact is obvious in the further examples.

3. The Hopf Bundle Geometry

EXAMPLE 4. The *Hopf bundle (fibration)*. This classical fibration can be found in many applications as a particular case of Example 3. Consider the sphere S^3 standardly embedded in the complex space $\mathbf{C}^2(z_1, z_2)$, as the set of points (z_1, z_2) such that $|z_1|^2 + |z_2|^2 = 1$.

Consider the two-dimensional sphere S^2 as the plane $\mathbf{R}^2$, completed with the point at infinity, and identify S^2 with the complex projective line $\mathbf{C}P^1$. Construct a mapping $p:S^3 \to S^2$ given by $p(z_1, z_2) = z_1/z_2$. It is easily seen that p is a smooth mapping regular at all points. Therefore, p defines a locally trivial fibration with the base S^2 and fiber S^1, since $p(e^{i\varphi}z_1, e^{i\varphi}z_2) = p(z_1, z_2)$. *The Hopf bundle can be described by representing the sphere S^3 as a union of two solid tori.* In fact, consider in C^2 a cone K given by the equation $|z_1| = |z_2|$. Its intersection with the sphere S^3 is homeomorphic to a torus $T^2 = S^1 \times S^1$. Therefore, the sphere is divided into the sum of two closed subsets Π_1 and Π_2, where $\Pi_1 = \{|z_1| \leqslant |z_2|\}$, $\Pi_2 = \{|z_1| \geqslant |z_2|\}$. They intersect along a two-dimensional torus. Each of the sets Π_i is homeomorphic to a solid torus $S^1 \times D^2$ (Fig. 25). The diagram is schematic in the sense that it shows the solid tori in $\mathbf{R}^3$, not in S^3. To make Fig. 25 exact, S^3 must be represented as $\mathbf{R}^3$ completed with the point at infinity (Fig. 26). In this representation, the solid

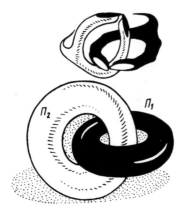

Figure 25

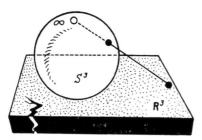

Figure 26

torus Π_1 can be considered standardly embedded in R^3 (Fig. 27) when its complement in $S^3 = \mathbf{R}^3 \cup \infty$ is homeomorphic to the second solid torus Π_2. If the standard parallels and meridians α_1, β_1 and α_2, β_2 (Fig. 28) are fixed on the solid tori, then pasting together the solid tori along their boundary is given by the diffeomorphism $h: T^2 \to T^2$ which sends parallels into meridians, and vice versa. The induced mapping of the one-dimensional homology groups $H_1 = \mathbf{Z} \oplus \mathbf{Z}$ of the torus is given by the integral matrix $h_* = \begin{pmatrix} 0 & 1 \\ -1 & 0 \end{pmatrix}$, $h(\alpha_1) = -\beta_2, h(\beta_1) = \alpha_2$. Now, we can describe the Hopf bundle geometrically. It suffices to exhibit a fibration of the sphere S^3 into circles. Fiber $\mathbf{R}^3 \cup \infty = S^3$ into a family of concentric tori (Fig. 29). To make it clear, we have cut $\mathbf{R}^3$ by the (z, x)-plane. There arise two limit situations, viz., the concentric tori contract to the standard circle in the (y, x)-plane (denote it by S_1^1; it is the dotted line in Fig. 29), and the tori, extending, converge to the critical axis Oz which is their "limit", since the whole infinity is glued into one point. The axis Oz represents a circle S_2^1 in the sphere S^3. It is clear that S_1^1 is the axis of the solid torus Π_1, S_2^1 is that of Π_2. To describe the fibers of the Hopf fibration, fix one of the tori which fiber $\mathbf{R}^3 \cup \infty$. Give on the torus a smooth circle

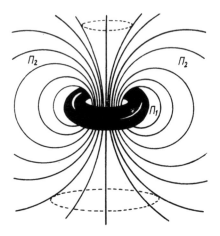

Figure 27

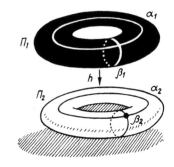

Figure 28

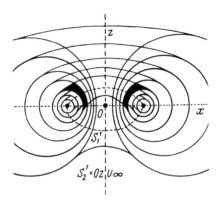

Figure 29

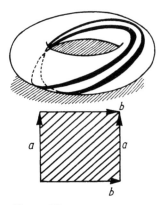

Figure 30

transversing it once along a parallel and meridian, respectively. Rotating this trajectory by orthogonal transformation which preserves the Oz-axis, we fiber the torus into the union of disjoint circles (Fig. 30). The same fibration is shown in Fig. 30 on the torus model, i.e., on a square with gluings on the boundary. Construct a similar fibration on each of the tori, the union of which is a three-dimensional sphere. Since they are disjoint, the circles fibering them do not intersect either. In both limit cases where the tori converge to S_1^1 and S_2^1, the circles fibering them converge to S_1^1 and S_2^1, respectively. This is the fibration S^3 giving the Hopf fibration (verify!). Any two fibers of this fiber space are interlinked, or a two-dimensional film, e.g., a disk, spanned over one of the fibers is sure to intersect another fiber. This is clearly seen in Fig. 30. (See also Fig. 31(a)). From the constructed mapping of Hopf fibration, we see the base points of the two-dimensional sphere S^2 into which some or other fiber-circles pass. Consider a disk D^2 in the (x, y)-plane with the boundary S_1^1. It is clear that any fiber (circle) different from S_1^1 must intersect the disk at one point exactly (Fig. 31(b)). When a fiber approaches the fiber S_1^1, its intersection point with the disk converges to the boundary of the disk, and when the fiber coincides with the fiber S_1^1, we must identify the circle $S_1^1 = \partial D^2$ with one point. The disk D^2 being transformed into a two-dimensional sphere, one and only one fiber (circle) of the Hopf fibration corresponds to each of the points. The fiber S_2^1 corresponds to the center of the disk (Fig. 31). Thus, to construct the Hopf fibration, we must fiber S^3 into the circles by the described method, and associate each circle with the point where it meets with the disk D^2 or the whole disk boundary if $S^1 = S_1^1$. The Hopf fibration is not trivial, otherwise (if $S^3 = S^1 \times S^2$) the group $H_1(S^3, \mathbf{Z})$ would be isomorphic to the group $\mathbf{Z}$ (prove!), which is impossible since $H_1(S^3, \mathbf{Z}) = 0$.

We dwell on this example, since it demonstrates many important effects characterizing generic fibrations. We give one more useful example of a fibration.

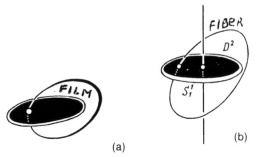

(a)

(b)

Figure 31

4. Geometry of the Fibration of Unit Tangent Vectors to a Sphere and the Index of a Vector Field

EXAMPLE 5. Let E be the space of all tangent unit vectors to an even-dimensional sphere S^{2n}, a base B a sphere S^{2n}, and projection $p:E \to B$ associate the tangent vector with its original point, i.e., the point of contingence to the sphere. This is a locally trivial fibration with a fiber S^{2n-1} (Fig. 32(a)).

LEMMA 1. *The fibration* $p:E \to S^{2n}$ *is non-trivial.*

PROOF. Assume the contrary, i.e., $S^{2n} \times S^{2n-1} = E$. The commutativity of the diagram from Definition 1 means that there is a non-zero cross-section of this fibration, i.e., a continuous mapping $h:S^{2n} \to E$ such that $ph = 1_{S^{2n}}$ (identity mapping of the base). But this would mean that, at each point $b \in S^{2n}$, a non-zero (unit) tangent vector $h(b)$ which is a continuous function of points b would be uniquely determined. It remains to prove that, on S^{2n}, there is no continuous tangent vector field different from zero at each point. This follows

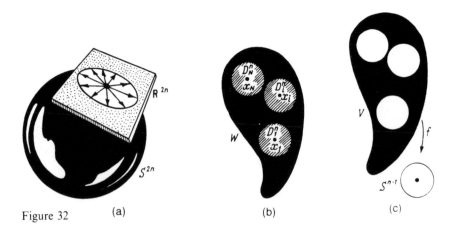

Figure 32   (a)   (b)   (c)

from the general theory of the index of a vector field on a sphere, which has other applications as well.

Recall the definition of the index of a vector field. Let $v(x)$ be a smooth field on a smooth manifold M^n, and the field have only a finite number of *singular points* x_0, i.e., where the field is zero. Then these points are isolated, i.e., for each of them, there exists an open neighborhood in which there are no other zeroes of the field. Consider in M^n a sphere S_ε^{n-1} of sufficiently small radius ε, with center at a singular point x_0. If ε is small, we can assume that $v(x) \neq 0$ on the whole sphere S_ε. Passing to local coordinates $x_1, \ldots, x_n$ in the neighborhood of x_0, we can assume that S_ε lies in a domain in R^n; therefore, we determine a mapping $f : S_\varepsilon \to S$, where $f(x) = \dfrac{v(x)}{|v(x)|}$ and $|v(x)|$ is Euclidean length $v(x)$.

DEFINITION 3 *The integer $\deg f$, i.e., the degree of the mapping $f : S_\varepsilon \to S$, is called the* **index** $\operatorname{ind}_{x_0} v$ *of a singular point x_0 of a vector field v. The sum of indices of singular points of the field for its all singular points is called the index of the vector field* $\operatorname{ind} v$.

It is known [1] that this definition is correct, and does not depend on the choice of local coordinates in the neighborhood of a singular point.

THEOREM 1. *Let $W^n \subset \mathbf{R}^n$ be a bounded domain in $\mathbf{R}^n$, the boundary ∂W of W being a smooth, compact and closed submanifold of dimension $n-1$ in R^n. Let v be a smooth vector field on W, with $v(x) \neq 0$ on ∂W and v having only finitely many singular points in W. Then the index of this field is equal to $\deg F$, where $F : \partial W \to S^{n-1}$ is a smooth mapping given by $F(x) = \dfrac{v(x)}{|v(x)|}$, where $x \in \partial W$.*

COROLLARY 1. *If two vector fields v_1 and v_2 on W are such that $\deg F_{v_1} = \deg F_{v_2}$ on ∂W, then their indices coincide. In particular, if v_2 is obtained from v_1 by a* **smooth homotopy** g_t *such that $g_t v_1(x) \neq 0$ on the boundary ∂W for $0 \leqslant t \leqslant 1$, then the indices of v_1 and v_2 coincide.*

Proof of Theorem 1. Let $x_1, \ldots, x_N$ be singular points of the field v, $N < \infty$. Enclose each singular point x_i by a sphere S_i^{n-1} of sufficiently small radius such that $D_i \cap D_j = \varnothing$ for $i \neq j$, where D_i is the n-ball bounded by the sphere S_i and having x_i as its center (Fig. 32(b)). Consider a domain V obtained from W by removing all balls D_i (Fig. 32(c)). Construct a smooth mapping $f : V \to S$ with $f(x) = \dfrac{v(x)}{|v(x)|}$, where S is the standard sphere of unit radius in $\mathbf{R}^n$. This mapping is correctly defined, since the field v is different from zero at all points of V. On the sphere S, consider the standard form of *Riemannian volume* ω, which is invariant under orthogonal transformations of $\mathbf{R}^n$. Then the integral $\int\limits_S \omega$ is equal to the volume of the sphere S, denoted by a_{n-1}. Being an exterior differential form on the sphere, the form ω induces a form $f^*\omega$ on V under the mapping f. Consider an integral $\int_V df^*\omega$, where d is the *exterior differential operation* [5, Ch. 6]. By the *Stokes formula* $\int_V df^*\omega = \int_{\partial V} f^*\omega$, where ∂V is the boundary of the domain V (sign is determined by the choice of orientation). But the boundary ∂V is decomposed into the disjoint sum of $(n-1)$-dimensional submanifolds $\partial W \cup (-\cup S_i)$. The sign $-$ indicates the orientation opposite to the induced. We obtain $\int_V df^*\omega = \int_W df^*\omega - \sum_i \int_{S_i} f^*\omega$. It follows from the properties of the operator d [5, Ch. 6] that it commutes with the form induction operation, i.e., $df^* = f^*d$. Thus, $\int_V df^*\omega = \int_V f^*d\omega$.

But the dimension of the form ω is maximal on the sphere S, i.e., $d\omega = 0$, or $\int_V df^*\omega = 0$. Thus, $\int_{\partial W} f^*\omega = \sum_i \int_{S_i} f^*\omega$. By Theorem 2 from [5, p. 394], we have $\int_{\partial W} f^*\omega = \deg f|_{\partial W} \cdot \int_S \omega = a_{n-1} \cdot \deg f|_{\partial W}$. Similarly, we obtain $\int_{S_i} f^*\omega = a_{n-1} \cdot \deg f|_{S_i}$. Thus, $a_{n-1} \cdot \deg f|_{\partial W} = a_{n-1} \times \sum_i \deg f|_{S_i}$. By definition of the index of a vector field, we obtain $\deg F = \sum_i \mathrm{ind}_{x_i} v = \mathrm{ind}\, v$. The theorem is proved.

THEOREM 2. *Any continuous vector field v on the sphere S^{2n} has necessarily at least one singular point x_0, i.e., such that $v(x_0) = 0$.*

PROOF. Assume the contrary, viz., that, on the sphere S^{2n}, there exists a tangent field different from zero at all points. By virtue of Theorem 1 from [1, p. 478], this field can be approximated as close as we please by a smooth tangent field denoted by the same symbol v. Since the original field is non-zero, the approximating smooth field is also different from zero at all points. We obtain Theorem 2 as a corollary of the following, more general, statement.

THEOREM 3. *Let v be a smooth **tangent vector field** on a sphere S^k having, probably, isolated singular points only. Then its index is equal to $1 + (-1)^k$, i.e., to zero on an odd-dimensional sphere, and to two on an even-dimensional sphere.*

Theorem 2 follows, since, assuming the contrary to the statement of the theorem, we have constructed on S^{2n} a tangent field without singularities, i.e., a field of zero index, which is impossible by Theorem 3.

Proof of Theorem 3. Let $x_0 \in S^k$ be an arbitrary point, and D_1 a disk in the sphere S^k of sufficiently small radius, having the point x_0 as its center (Fig. 33(a)). Represent S^k as the union of two closed disks $D_1 \cup D_2$, where D_2 is the closure of the complement to D_1 in S^k. Let x_0 be a non-singular point in the field v, i.e., $v(x_0) \neq 0$. We can assume that, on the whole disk D_1, the field is different from zero, and, moreover, that it can be made arbitrarily close to a "parallel field", i.e., the integral trajectories of v can be arbitrarily near to the pencil of parallel lines (Fig. 33(b)). Since the field does not have singular points

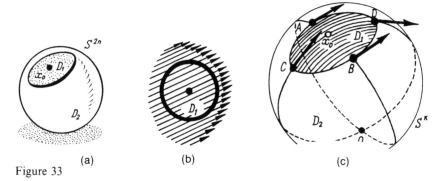

Figure 33 (a) (b) (c)

on D_1, all singular points (if they exist) are concentrated on the disk D_2. This disk can be translated by a developing diffeomorphism (e.g., by the standard stereographic projection with center at the point x_0) onto the standard disk in $\mathbf{R}^k$ with center at the point 0. In Fig. 33(c), this operation is equivalent to the following: D_1 must be removed from the sphere, and the remaining portion developed and placed on the plane. Observe how the vectors of the field v are shown on the boundary of the disk D_2 after this operation, i.e., after introducing Euclidean coordinates on D_2. It is clear that, from the stand point of these coordinates in the disk D_2, the field on its boundary assumes the form shown in Fig. 34(a). Even now, it is intuitively obvious that, in the two-dimensional case, e.g., the presence of this turbulence of the field on the disk boundary means that there is at least one singularity inside the disk. Now, we establish this fact on the basis of Theorem 1. The disk $D_2 \subset \mathbf{R}^k$ can be identified with the domain W from Theorem 1. Then the boundary ∂W is the sphere S^{k-1}, and the field v on ∂W is constructed as shown in Fig. 34(a) for the two-dimensional case. It follows from Theorem 1 that, to find the index of a field on the disk D_2, it is sufficient to compute the degree of a mapping $F:S^{k-1} \to S^{k-1}$, where $F(x) = v(x)/|v(x)|$. To this end, it suffices to consider a regular point, and compute the signs of the Jacobians at all its inverse images. Since the degree of a mapping does not depend on the choice of a regular point, we can for convenience take a point C on the sphere S^{k-1}, while identifying two copies of S^{k-1} and assuming that the mapping F carries the sphere into itself (Fig. 34(a)). It follows from the construction of v that two points only are the inverse images of the point C: the point C itself and its diametrically opposite point D. The mapping F at the point C is shown in Fig. 34(b). Its differential defined on the tangent plane to the sphere at C is given by a diagonal matrix with positive numbers on the diagonal. Therefore, the sign of the Jacobian at the point C is plus. At the point D, the situation is different. Each vector v close to the point D is mapped into a vector v' emanating from the point O and obtained from v by a translation to O (Fig. 34(c)). Hence, the differential dF at D is denoted on the tangent plane to the sphere by a diagonal

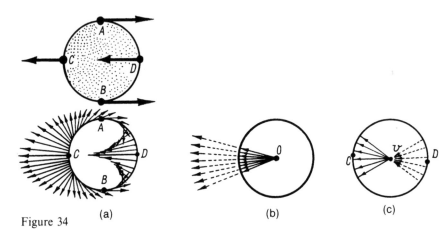

Figure 34   (a)   (b)   (c)

negative numbers on the diagonal. Thus, we deal with a reflection of the sphere in its center. The degree of the mapping $\alpha(x) = -x$ of the sphere S^{k-1} onto itself equals $(-1)^k$. Therefore, the sign of the Jacobian at D is $(-1)^k$, which proves Theorem 3.

In popular scientific literature, Theorem 3 is sometimes referred to as the Poincaré-Brouwer theorem: The "hedge-hog", which has rolled itself up into a ball and cannot be combed so that its spines lie along the tangent (if k is even).

Thus, the tangent bundle $p: E \to S^{2n}$ is not trivial. The fibration $p: E \to S^1$ is trivial, since it is homeomorphic to $S^1 \times S^0$, where S^0 is a zero-dimensional sphere, a pair of points.

Above, we have made a detailed study of some properties of the Hopf fibration, since the latter is used in the sequel. Let us point out one curious property more. If the homotopy groups of a finite cell complex are trivial, then it contracts to a point (if the cell complex is connected). If the complex is simply-connected, then the same fact follows from the triviality of integral homology groups. The question arises: will a continuous mapping of one complex into another be homotopic to a mapping onto a point if the former induces the zero homomorphism of all homotopy (or homology) groups? It turns out that it will not. Consider a Hopf fibration $p: S^3 \to S^2$. Let $T^3 = S^1 \times S^1 \times S^1$ be a three-dimensional torus. Contracting its two-dimensional skeleton to a point, we obtain a continuous mapping $g: T^3 \to S^3$. Taking a composition $f = pg$, we obtain a mapping $f: T^3 \to S^2$ which, as is easily verified, is not homotopic to a mapping into a point. In fact, if the mapping f were homotopic to zero, then, by the covering homotopy theorem, this homotopy could be covered by some homotopy in S^3. Consequently, we would obtain that the mapping g is homotopic to the mapping carrying the torus into a fiber of the Hopf fibration, i.e., into a circle. Since this circle is contracted along the sphere S^3 to a point, the mapping g would be homotopic to zero, which contradicts its construction. Meanwhile, it is easily seen that induced mappings of the integral homology and cohomology are trivial. Moreover, the induced mappings of the homotopy groups $f_* : \pi_i(T^3) \to \pi_i(S^2)$ are trivial, since the two-dimensional sphere is simply connected, and the homotopy groups of the torus T^3 are equal to zero, beginning with dimension two. A similar example can be constructed in the case where both complexes are simply connected.

CHAPTER II/FUNCTIONS ON MANIFOLDS

1. Exact Morse Functions

The present section is to some extent based on the material given in [1, 3, 34], and it is an immediate sequel of a section in [34], dealing with the study of Morse functions on a manifold.

Recall briefly that a *Morse function* is characterized by the fact that its critical points are non-degenerate. A point x of a function f is said to be non-degenerate if, in the vicinity of x, there exists a regular system of coordinates in which f is a diagonal quadratic form of all indeterminates. The number of minuses of a diagonal quadratic form is said to be *the index of a critical point*. A Morse function in general position is one assuming various values at critical points. In the sequel, we are going to discuss Morse functions in general position. Recall that a Morse function is said to be *regular* if, for any two critical points x of index λ and y of $\lambda + 1$, a strict inequality $f(x) < f(y)$ holds. Denote by F_λ the set of all Morse functions with minimum (for the given set) critical points of index λ.

DEFINITION 1. A function with minimum critical points in each dimension, i.e., $f \in \bigcap_{\lambda=0}^{n} F_\lambda$, is called *an exact, or minimal, function on the manifold M^n*.

In other words, for all λ, the number of critical points of index λ of a Morse function is the least possible in the class of all Morse functions on the given manifold. If $N_\lambda(f)$ is the number of critical points of index λ, then, for the exact Morse function f, an equality $N_\lambda(f) = \min N_\lambda(g)$ holds, where minimum is taken over all Morse functions g on the manifold. Generally speaking, an exact Morse function on the given manifold may be non-existent. It turns out that if the manifold is not simply connected, then there exist topological invariants which, in some cases, obstruct the existence of even one exact Morse function on a manifold M (in the sense of Definition 1), whereas an exact Morse function always exists for simply connected manifolds of sufficiently large dimension (see below).

The problem of the existence of an exact Morse function arises naturally when the topological structure of a given manifold is analyzed. Recall that each Morse function on M is naturally connected with some decomposition of M in the union of handles [2, 3, 34]. In turn, such representation enables us to describe the homotopy structure of a manifold in the form of a cell complex in which the number of cells of dimension λ is equal to the number of critical points of index λ of a Morse function. Different Morse functions correspond,

generally speaking, to different decompositions of one and the same manifold in the sum of handles. By minimizing the number of critical points, we minimize the number of handles, i.e., we obtain a simpler representation of the manifold in the form of the "sum of elementary blocks". Therefore, the problem of finding a Morse function with minimum number of critical points arises from the problem of constructing cellular decomposition of a manifold in the sum of a minimal number of cells.

Consider the representation of the Abelian homology group $H_\lambda(M^n, \mathbf{Z})$ as the direct sum of a free Abelian group G_λ and a finite group K_λ. Denote by p^λ the rank of the group G_λ, i.e., the minimum number of generators in the group G_λ, and by q^λ the minimum number of generators of the group K_λ.

THEOREM 1 (see [51]). *Let M^n be a connected smooth closed manifold of dimension $n \geqslant 6$. Then on M^n there exists an exact Morse function, a number N_λ of the critical points of the index λ being computed in the following way:* $N_\lambda = p^\lambda + q^\lambda + q^{\lambda-1}$.

PROOF. We give the proof of this theorem in a nutshell, recommending [51] or [53] to the reader interested in the details. Let $f: M^n \to I$ be a regular Morse function on the manifold M, where I is a unit interval. In [3], see the proof of the existence of regular functions. Let c_λ be such numbers that $f(x) < c_\lambda < f(y)$ for any critical points x and y of indices λ and $\lambda+1$, respectively. The level surface $\{f = c_\lambda\}$ separates x and y from each other. By M_λ denote the following manifold (with boundary) $M_\lambda = f^{-1}[0, c_\lambda]$ (Fig. 1). Let $\xi(f)$ be the gradient-like vector field for the function f [3, 34]. The specification of the function f and the field $\xi(f)$ determines the decomposition of the manifold into the sum of *handles* $D^\lambda \times D^{n-\lambda}$ [3, 34]. We obtain the sequence of inclusions $M_0 \subset M_1 \subset \ldots \subset M_n = M^n$. It is generally known [3, 34] that

$$H_i(M_\lambda, M_{\lambda-1}, \mathbf{Z}) = \begin{cases} 0 & \text{for } i \neq \lambda, \\ \mathbf{Z} \oplus \ldots \oplus \mathbf{Z} \ (k \text{ summands}) & \text{for } i = \lambda, \end{cases}$$

where k is the number of critical points (handles) of index λ. The basis for the

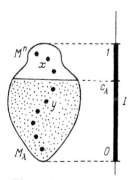

Figure 1

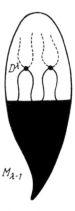

Figure 2

group $H_i(M_\lambda, M_{\lambda-1}, \mathbf{Z})$ is given by the handles of index λ with fixed orientations (Fig. 2). Geometrically, the homology groups generators can be given in the form of separatrix disks D^λ of dimension λ, which consists of separatrices of the field $\xi(f)$ forming critical points of index λ. Considering these disks by the modulus of the subspace $M_{\lambda-1}$, we obtain relative cycles realizing the basis for the relative homology group. Consider a chain complex $\{C, \partial\}$ composed of groups $C_\lambda = H_\lambda(M_\lambda, M_{\lambda-1}, \mathbf{Z})$ and connecting homomorphisms ∂_λ, where ∂_λ is the boundary operator from the exact sequence of the following triple $M_\lambda \supset M_{\lambda-1} \supset M_{\lambda-2}$, viz., $\partial_\lambda : H_\lambda(M_\lambda, M_{\lambda-1}, \mathbf{Z}) \to H_{\lambda-1}(M_{\lambda-1}, M_{\lambda-2}, \mathbf{Z})$.

LEMMA 1. *Homology groups of a chain complex $\{C, \partial\}$, i.e., the group $H_\lambda(C, \partial) = \operatorname{Ker} \partial_\lambda / \operatorname{Im} \partial_{\lambda+1}$, are canonically isomorphic to the homology groups of the manifold $H_\lambda(M, \mathbf{Z})$.*

For the proof of the lemma see, e.g., [3]. This lemma enables us to compute the homology of a manifold, proceeding from the knowledge how the handles are situated. It is known (see [3]) that the operator ∂_λ is given by the matrix of the intersection numbers of the right-hand spheres and the left-hand ones which are the boundaries of the corresponding separatrix dicks of indices λ and $\lambda+1$.

Next is what can be called the lemma on the cancellation of handles of nearby indices. It turns out that, in some cases, the handles nearby indices λ and $\lambda+1$ can "eat up" each other, which is equivalent to cancelling two critical points of such indices. As a result, we obtain a Morse function with a smaller number of critical points. In particular, this operation enables us to reduce the number of handles in the manifold decomposition into the sum of handles.

LEMMA 2 (see [54]). *Let $M' = M_{\lambda-1} \cup_\varphi D^\lambda \times D^{n-\lambda} \cup_\psi D^{\lambda+1} \times D^{n-\lambda-1}$, where $D^\lambda \times D^{n-\lambda}$ and $D^{\lambda+1} \times D^{n-\lambda-1}$ are handles of indices λ and $\lambda+1$, respectively, glued to the manifold $M_{\lambda-1}$ with respect to the mappings φ and ψ. Let*

$\pi_1(M^n) = 0$, $\lambda \geq 2$, $\lambda + 1 \leq n - 3$, $n \geq 6$. *Assume, too, that the intersection number of the left-hand sphere of the handle $D^\lambda \times D^{n-\lambda}$ with the right-hand sphere of the handle $D^{\lambda+1} \times D^{n-\lambda-1}$ is equal to ± 1. Then, there exists a diffeomorphism $h: M \to M$, identity outside some open neighborhood of the manifold M' in M such that $h(M') = M_{\lambda-1}$. In other words, two handles have cancelled out.*

Recall the process of adding handles. Consider a manifold $W' = M_{\lambda-1} \cup_\varphi D^\lambda \times D^{n-\lambda}$ obtained from the manifold $M_{\lambda-1}$ by gluing the handle of index λ. Assume that M^n is simply connected and $\lambda \geq 2$. Then a mapping $\varphi|_{\partial D^\lambda \times 0}$ defines an unambiguously defined element $[\varphi]$ in the group $\pi_{\lambda-1}(\partial M_{\lambda-1})$.

LEMMA 3 (see [54]). *Let $M' = M_{\lambda-1} \cup_\varphi D^\lambda \times D^{n-\lambda} \cup_\psi D^{\lambda+1} \times D^{n-\lambda-1}$, where $\lambda \geq 2$, $n - \lambda \geq 2$. There exists such a mapping $\tilde{\psi}: \partial D^\lambda \times D^{n-\lambda} \to \partial M_{\lambda-1}$ isotopic to the embedding ψ that $[\tilde{\psi}] = [\varphi] + [\psi]$. We assume that the images of the mappings φ and ψ do not intersect. Similarly, there exists such a mapping $\tilde{\psi}$ that $[\tilde{\psi}] = [\psi] - [\varphi]$. Therefore, intersection numbers of a new handle constructed from the embedding $\tilde{\psi}$ are obtained by adding (resp. subtracting) the original indices. Under the operation of adding handles, the matrix of the homomorphism $\partial_\lambda: H_\lambda(M_\lambda, M_{\lambda-1}) \to H_{\lambda-1}(M_{\lambda-1}, M_{\lambda-2})$ is subject to an elementary transformation.*

Now, we turn to the proof itself of Theorem 1. Let f be some regular Morse function on M^n, and $\xi[f]$ the gradient-like vector field. Construct the corresponding manifold decomposition into the sum of handles. From the condition of Theorem 1, it follows that, in this decomposition, there are no handles of indices 1, n-1 and there is only one handle of index 0 (minimum of the function) and of index n (maximum of the function), respectively. Consider a chain of inclusions $D^n = M_0 \subset M_2 \subset M_3 \subset \ldots \subset M_{n-2} \subset M_n = M^n$. Construct a corresponding chain complex $\{C, \partial\}$ and, on the basis of Lemma 1, obtain that $H_i(C, \partial) = H_i(M^n, \mathbf{Z})$. Consider the handles of indices 2 and 3 and a homomorphism $\partial_3: H_3(M_3, M_2, \mathbf{Z}) \to H_2(M_2, M_0, \mathbf{Z})$, which is given by the matrix of intersection numbers of the corresponding right-hand spheres and left-hand ones. Using the process of adding handles, described in Lemma 3, reduce the matrix of the homomorphism ∂_3 to a diagonal form

$$\begin{pmatrix} \varepsilon_1 & \cdot & \cdot & & & & \\ & & \cdot & \varepsilon_k & & 0 & \\ & & & & n_1 & \cdot & \cdot \\ & & 0 & & & & \cdot & n_l \end{pmatrix}$$

where $\varepsilon_i = \pm 1$, n_i divides n_{i+1}, $n_i \in \mathbf{Z}$.

It is clear that the handles with the intersection number of spheres equal to ε_i do not contribute to the homology group, and, on the basis of Lemma 2, these pairs of handles can be cancelled. There remain the handles of index 2, which do not intersect the handles of index 3 and correspond to the free generators of the group $H_2(M^n, \mathbf{Z})$ as well as the handles of indices 2 and 3 which intersect with the indices equal to n_i, which corresponds to the finite order elements in the group $H_2(M^n, \mathbf{Z})$. Using a dual partition into handles (i.e., the function $1 - f$), reduce similarly the handles of indices $n - 2$ and $n - 3$. Consider then a kernel of the homomorphism ∂_3. It is a free Abelian group generated by the handles of index 3, which do not intersect the handles of index 2. Eliminiate extra handles of index 3 by means of the handles of index 4. We obtain that $\partial_4 H_4(M_4, M_3, \mathbf{Z}) \subset \operatorname{Ker} \partial_3$. Reduce the matrix of the homomorphism ∂_4 to a diagonal form and, by means of similar considerations, prove that there exists $p^3 + q^3$ of non-removable handles. Thus, after two steps, we have obtained a decomposition of the manifold M^n, having $p^2 + q^2$ of handles of index 2 and $p^3 + q^3 + q^2$ of handles of index 3. Iterating this process, we can consecutively eliminate extra handles of the remaining indices. The theorem is proved.

THEOREM 2 (see [53]). *Let M^n be a simply connected smooth closed manifold of dimension $n \geqslant 6$. Let f and g be two exact Morse functions on a manifold M. Then, there exists an effectively computable obstruction to the existence of a diffeomorphism $h : M \to M$, such that $f = gh$.*

The condition of simple connectedness of the manifold M^n is essential for the validity of Theorems 1 and 2. Simple examples show that, in the non-simply connected case, the minimum number of critical points of a Morse function on M^n depends, at least, on the homology of a universal covering manifold $\tilde{M}^n$ and some fine topological invariants. Furthermore, it turns out (see [53]) that there arises some non-uniqueness in the construction of exact Morse functions (in the cases where they exist); see [52]. In some cases, nevertheless, we succeed in proving the existence of an exact Morse function and computing a number of its critical points in each dimension. We give only some of these results here. A full survey see in [53].

THEOREM 3 (see [53]). *Let M^{n+k} be a smooth manifold of dimension $n \geqslant 6$, and let $k \geqslant 0$, $\pi_1(M^{n+k}) = \mathbf{Z} \oplus \ldots \oplus \mathbf{Z}$ (k times). Then, on the manifold M^{n+k}, there exists an exact Morse function. In addition, the number N_λ of critical points of index λ of this function is equal to $p^{\lambda-k}(1+p)^k + q^{\lambda-k-1}(1+q)^{k+1}$, where p^i, q^i (after brackets are removed in accordance with usual algebraic laws) are considered to be the minimum number of free generators in the homology group $H_i(\tilde{M}^n, \mathbf{Z})$ and the minimum number of generators in a finite-order subgroup in the group $H_i(\tilde{M}, \mathbf{Z})$, respectively. Here, $\tilde{M}^n$ is a universal covering manifold for the manifold M^n.*

For $k = 0$, we obtain Theorem 1 (due to Smale). When $k = 1$, we get $N_\lambda = p^\lambda + p^{\lambda-1} + q^\lambda + 2q^{\lambda-1} + q^{\lambda-2}$. When $k = 2$, we have $N_\lambda = p^\lambda + 2p^{\lambda-1} + p^{\lambda-2} + q^\lambda + 3q^{\lambda-1} + 3q^{\lambda-2} + q^{\lambda-3}$, etc. Thus, with the growth of the fundamental group of the manifold I, homology groups of a still wider range of dimensions take part in forming the minimum number of critical points of given index λ.

2. Multidimensional Analog of the Morse Theory

This section is dedicated to the statement of some aspects of the Morse theory analog for many-valued functions. First, we dip briefly into a new theory constructed by S. P. Novikov in [59, 60, 61].

Let M be a certain manifold, finite-dimensional or infinite-dimensional, on which an exterior closed 1-form ω is given, i.e., $d\omega = 0$. Along the paths in the manifold M, integrals of this form define a *"many-valued functional"* S on the set of all paths. Naturally, such functionals appear in physical problems; therefore, they arouse interest by themselves [60]. In particular, the following question arises immediately: Can an analog of Morse theory for many-valued functionals (functions) be constructed, i.e., can a relation of the numbers of stationary (critical) points (where $dS = 0$) of various indices to the topology of the manifold M be found?

Natural examples of many-valued functionals were considered in [61], where periodic solutions of Kirchhoff type equations, a spinning top in a gravitation field and the like were studied. The most general class of examples of many-valued functionals may be determined as follows. Consider two smooth manifolds N^q and M^n where N^q is supposed to be compact, and let S_0 be a functional on the space F of smooth mappings $f : N \to M$. By open sets U_i, specify on M^n a closed $(q + 1)$-form ω and the covering $M^n = \cup_i U_i$ of power of continuum such that the following properties are fulfilled:

(i) For any mapping $f : N \to M$, the image of manifold N, i.e., $f(N)$ is wholly in a set U_i. The index i depends, generally speaking, on the mapping f;
(ii) The form ω is exact on the set U_i, i.e., $\omega = d\psi_i$.

Then, naturally, there arises a certain many-valued functional S on an infinite-dimensional space F of mappings $f : N \to M$, viz., for any mapping $f : N \to U_i$, put

$$S_i(f) = S_0(f) + \int_{f(N)} \psi_i. \qquad (*)$$

Then, the functional S_i can be considered to be a "branch" of a many-valued functional S defined on an open subset in the mapping space. The functionals S_i are, in this sense, the "local functionals". Gluing them, we can construct a globally definite but, generally speaking, many-valued functional S.

If we consider an intersection of the sets $U_i \cap U_j$, and if the image $f(N)$ is wholly contained in this intersection, then $S_i(f) - S_j(f) = \int_{f(N)} (\psi_i - \psi_j)$, where $d(\psi_i - \psi_j) = 0$, i.e., $\psi_i - \psi_j$ is a closed q-form on the intersection $U_i \cap U_j$. The latter means, in fact, that a 1-form δS is correctly defined on the whole infinite-dimensional functional manifold F. In fact, $\delta(S_i - S_j)(f) = \int d(\psi_i - \psi_j) = 0$, i.e., the 1-form δS is computed similarly both in the "chart" U_i and in the "chart" U_j. Further, it is clear that the many-valuedness of the functional S on the space F has its origin in a possible non-homotopy of various paths connecting two fixed points in the manifold F.

Therefore, passing from F to a cofering manifold $\tilde{F}$ (its fundamental group is "smaller" than that of F), we obtain on the covering $\tilde{F}$ a 1-form which is the differential of a single-valued function on the covering $\tilde{F}$. Summing up, we formulate the following lemma.

LEMMA 1 (see [60]). *The set of local functionals $S_i(f)$ determines a closed 1-form δS on the function space F of admissible mappings $f : N \to M$. This 1-form δS determines the many-valued functional S on the mapping space F. This functional is single-valued on some infinite dimensional and infinite-sheeted covering $\tilde{F} \to F$.*

Consider the following example. Let $N^q = S^1$ be a circle, then an infinite-dimensional manifold F is the space of all closed paths (with a parametrization) in a manifold M. If the manifold M is simply connected, then the space F is connected. As an initial functional S_0, we can consider a usual length functional defined on the paths to M, i.e., on the mapping space $S^1 \to M$. This functional generates a functional S which is, generally speaking, many-valued (see above).

THEOREM 1 (see [59]). *Let M^n be a simply connected complete Riemannian manifold (i.e., with positive definite metric) and S_0 a length functional defined on the space F of the closed paths in the manifold M. Then, for any closed 1-form ω, on the manifold M^n a functional S of the form (*), single-valued or many-valued, has at least one periodic extremal (a critical point of the functional on the manifold F).*

Now, we turn to the question of a manifold homology connection with a number of critical points of many-valued functions. Let M^n be a finite-dimensional manifold and ω a closed 1-form on M. This form defines an element Ω of a cohomology group $H^1(M, \mathbf{R})$. Since $H^1(M, \mathbf{R}) \approx H_1(M, \mathbf{R})$, the element Ω determines a one-dimensional cycle with decomposition real coefficients in terms of an integral basis for the group $H_1(M, \mathbf{Z}) \otimes \mathbf{R} = H_1(M, \mathbf{R})$. Hence, there exists a covering of infinitely many sheets $p : \tilde{M} \to M$ such that the form $p^*\omega$ is exact on $\tilde{M}$, i.e., the differential of a function S on this covering $p^*\omega = dS$. The simplest example is the standard 1-form $\omega = d\varphi$ on a plane with an exclusion of the coordinate origin, then the covering p is a Riemannian logarithm surface. The function S correctly defined on the covering is referred to as a "many-valued function" on the original manifold M. Further, we shall suppose that all its critical points are either non-degenerate or they generate non-degenerate critical submanifolds in the underlying manifold. We also suppose that, for the functional S, the steepest descent is defined correctly, i.e., on the manifold M, any compact set while descending along the integral field trajectories of grad S either "hangs" from some critical point (i.e., the deforming set "brings" a critical point along with some of its points) or, moving down, passes through all surfaces of the level S

consecutively. Denote by $N_\lambda(S)$ (or by $N_\lambda(\omega)$) a number of critical points with Morse index λ of the function S, where $p^*\omega = dS$. Just as in the case of single-valued functions, one may query how to estimate the number of critical points of the many-valued function S (i.e., a closed 1-form ω) of index λ with the help of the topological invariants of the manifold.

In the group $H_1(M, \mathbf{Z})$, we can choose a basis $(\gamma_1, \ldots, \gamma_k, \gamma_{k+1}, \ldots, \gamma_q)$ such that, for integrals of the form ω on one-dimensional cycles γ_i, the following equalities hold:

$$\int_{\gamma_j} \omega = \begin{cases} 0 & \text{if } j \geqslant k+1 \\ \chi_j \neq 0 & \text{if } j \leqslant k \end{cases},$$

no linear combination of all numbers χ_j for $j = 1, \ldots, k$ being zero with rational (or integral) coefficients. The number $k - 1$ is called the irrationality degree of the form ω. A monodromy group (group of shifts) of a minimal covering $p : \tilde{M} \to M$ on which the inverse image of ω transforms into a differential of a single-valued function $p^*\omega = dS$ is isomorphic to the group $\mathbf{Z}^k$. This group is free Abelian, and its generators $t_1, \ldots, t_k$ are realized as translations on the covering $\tilde{M}$, i.e., $t_j : \tilde{M} \to M$. The simplest case, and interesting at that, is where $k = 1$, when the form ω (possibly after being multiplied by some real number) coincides with an element of the integral cohomology group $[\omega] \in H^1(\mathbf{M}, \mathbf{Z})$. In our case, the value $(2\pi i S)$ is a single-valued function; hence, a single-valued mapping $f = \exp(2\pi i S) : M^n \to S^1$ is correctly defined. The properties of critical points of such mappings and their relation to the manifold topology are studied in [60].

It is obvious that if the mapping f constructed above has no critical points, then f defines a locally trivial fibration $f : M^n \to S^1$ with the base $B = S^1$. Since $k = 1$, the covering $p : \tilde{M} \to M$ is $\mathbf{Z}$-cyclic, i.e., its fiber bundle is a group of integers $\mathbf{Z}$. We describe the process of constructing this covering in detail. Let D be the Poincaré duality operator establishing an isomorphism between the groups $H^1(M^n, \mathbf{Z})$ and $H_{n-1}(M^n, \mathbf{Z})$. Applying it to a one-dimensional cocycle $[\omega]$, we obtain an element $D[\omega] \in H_{n-1}(M^n, \mathbf{Z})$. Realize the cycle $D[\omega]$ by a submanifold K^{n-1} in M^n. Cutting the manifold M^n along the cycle-submanifold K^{n-1}, we obtain a film W^n with two boundaries $\partial W = K_0^{n-1} \cup K_1^{n-1}$ diffeomorphic to K^{n-1}. Take infinitely many copies of the film $W_i \approx W$ with boundaries $\partial W_i = K_{i,0} \cup K_{i,1}$, where $K_{i,\alpha}$ are diffeomorphic to K^{n-1}. Paste them together along the boundaries according to the rule indicated by the following numeration: $\tilde{M} = \cup_i W_i$, $K_{i+1,0} \approx K_{i,1}$, $-\infty < i < +\infty$ (Fig. 3). As a result, we obtain a "tube infinite in both directions" composed of an infinite number of copies of the film W, which are pasted together. It is also clear that the function S is correctly defined on the tube $\tilde{M}$, $\tilde{M}$ being a covering of the original manifold M. Forming this covering has destroyed the cocycle $[\omega]$. Consider now the level surfaces of the function S. By choice of the submanifold K realizing the cycle $D[\omega]$ we may assume that $K = K_{0,0}$ is one of

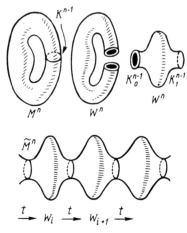

Figure 3

the level surfaces of the function S, i.e., the full inverse image of the point under the mapping $f = \exp(2\pi i S)$ of the manifold $\tilde{M}$ into a circle. It follows from the definition of the covering that the operator t of the monodromy acts on the covering in the following way: $t : W_i \to W_{i+1}$, $K_{i,0} \to K_{i,1} = K_{i+1,0}$ (Fig. 3).

According to classical Morse theory, a smooth Morse function given on the finite-dimensional manifold M generates, under some natural conditions, a cell complex which is a decomposition of $\tilde{M}$ in the sum of cells. However, in the case under discussion, an important condition necessary for the validity of general Morse theory is not fulfilled; viz., to correctly realize Morse surgeries, each closed domain of smaller values $\{S \leqslant a\}$ (for all a) should be relatively compact. Only when this condition is fulfilled, the domain of values $\{S \leqslant a + \varepsilon\}$, where a is a certain critical value for the function S, can be obtained from the domain $\{S \leqslant a - \varepsilon\}$ by gluing a certain number of handles. In our case, domains of smaller values are not relatively compact. However, from each critical point of index λ, (for S) the "steepest descent" surface emanates, an analog of separatrix disk of dimension λ in the classical case. This surface or its sufficiently small perturbation, if necessary, can be considered to be a "cell" of dimension λ.

However, in contrast to the classical case, this "cell" can extend along the levels of the function S to $-\infty$, and, therefore, an infinite number of such "cells" of dimension $\lambda - 1$, corresponding to critical points of index $\lambda - 1$, are contained in the algebraic boundary of the "cell".

Under $t : \tilde{M} \to \tilde{M}$, the function S is carried into itself with a certain additive constant. Meanwhile, critical points of S are sent into critical points again. Therefore, we obtain:

(i) Each critical point of S determines a free generator realized by an "infinite cell" whose dimension is equal to the index of the critical point in the chain complex; (ii) The boundary of each such cell is a linear combination (possibly, infinite) of cells of this cell complex, lying "below" the levels of function S, i.e., tending to infinity in one direction only in the covering ("tube") $\tilde{M}$; (iii) All cells of the constructed complex are obtained from a finite number of basic cells by all sorts of displacements to the elements t^m of the group $\mathbf{Z}$ acting on the covering $\tilde{M}$ as described above.

Describe an algebraic construction on the basis of which, as it turns out (see [59, 60]), we can obtain Morse type inequalities for many-valued functions. Consider a ring L, composed of the Laurent series of the following form: $\sum_{j > \text{const} > -\infty} m_j t^j$, where m_j are the integers vanishing for all sufficiently large negative subscripts j. A cell complex generated by a many-valued function on a manifold M or by a single-valued function S on its covering $\tilde{M}$ can be considered as a free complex C consisting of L-modules C_i with a finite number of generators. Here we make use of the fact that a number of critical points on each manifold W_i is finite. The complex C has the following form:

$$0 \to C_n \xrightarrow{\partial} C_{n-1} \xrightarrow{\partial} \ldots \xrightarrow{\partial} C_1 \xrightarrow{\partial} C_0 \to 0,$$

where the boundary operator ∂ is a homomorphism of L-modules. In contrast to the general Morse theory, a situation is possible when $C_0 = C_n = 0$. When a mapping $f : M^n \to S^i$ determines a locally trivial fibration with a base circle, there exists a 1-form ω on M, having no critical points at all. The inverse image of the form $d\varphi$ on S^1 can be taken as such a form. In this case, the complex is trivial, i.e., $C_i = 0$ for all i. Nevertheless, just as in the classical case, the complex C admits computing homological characteristics of the manifold.

LEMMA 2 (see [60]). *The homology of the complex C of L modules, generated by any smooth closed 1-form ω without non-degenerate critical points, and by the correctly defined steepest descent are homotopy invariant.*

Therefore, the homology group invariants of the described complex can be used to obtain Morse inequality analogs in the case of many-valued functions generating mappings (see above) into the circle $f = \exp(2\pi i S) : M^n \to S^1$. It is easily seen that the submodules of the free modules C_i are free. Hence, in the modules of the cycles $Z_i = \text{Ker}\,\partial \subset C_i$ and boundaries $B_i = \text{Im}\,\partial \subset C_i$, free bases can be chosen. The difference of the ranks A and B of these free submodules is said to be the *Betti number*; it is denoted by $p^i(M^n, a)$, where $a = [\omega] \in H^1(M^n, \mathbf{Z})$. It is clear that, in the classical case, these numbers are transformed into ordinary Betti numbers $p^i(M^n)$ (see Section 1 of the present chapter). The analogs $q^i(M^n, a)$ of the classical coefficients of torsion $q^i(M^n)$ are defined as follows. Choose a free basis $e_1, \ldots, e_A$ in the module of the cycles Z_i, and a free basis $e'_1, \ldots, e'_B$, where $A - B = p^i$, in the submodule of the boundaries B_i. This can always be done so that

(i) $e'_j = (n_j + \sum_{k \geqslant 1} n_{jk} t^k) e_j + \sum_{i > B} q_{ij}(t) e_i;$

(ii) The number n_j is divisible by the number n_{j+1};

(iii) The exponents of all terms of the series $q_{ij}(t)$ are non-negative;

(iv) The numbers $q_{ij}(0) \neq 0$, and are divisible by n_j for all i, i (if the series does not vanish identically).

The total number of indices j such that $n_j \neq 1$ is called a *torsion coefficient* denoted by $q^i(M^n, a)$. The number $p^i + q^i$ coincides with the minimum number of generators of the module $H_i = Z_i / B_i$, which is the natural analog of the classical case.

THEOREM 2 (see [60]). *The analogs of the* **Morse inequality** *are valid for the numbers* $N_\lambda(S)$ *(or* $N_\lambda(\omega)$*) of critical points of the index* λ *for the mapping* $f = \exp(2\pi i S)$ *of the manifold* M^n *into the circle* S^1 *or the closed 1-form* ω*, where* $a = [\omega] \in H^1(M^n, \mathbf{Z})$*, i.e., the form* ω *coincides with some element of an integral one-dimensional cohomology group. These inequality analogs have the form*

$$N_\lambda(S) \geqslant p^\lambda(M^n, a) + q^\lambda(M^n, a) + q^{\lambda - 1}(M^n, a).$$

In form, these inequalities coincide with the classical (see the preceeding section), though the topological invariants involved have a deeper and more complicated geometrical meaning.

If the manifold M^n has the fundamental group isomorphic to the group $\mathbf{Z}$ and if the 1-form ω realizes a cocycle corresponding to a generator of this group, then the covering $\tilde{M}^n$ is simply connected and, therefore, a question arises about the accuracy of Morse inequality analogs obtained in Theorem 2 in the same sense as is understood in the classical case (see the preceeding section for the Smale theorem). In other words, one may query as to whether it is possible to construct a many-valued function for transforming the inequality in Theorem 2 into equality, i.e., whether an exact many-valued function exists always. To construct a "minimal" (in this sense) 1-form ω, a choice of a "minimal", in some sense, initial submanifold $K^{n-1} \subset M^n$ realizing the cycle dual to the cocycle ω is essential. It turns out that the exact many-valued Morse function always exists for the above-mentioned case as proved by M. Farber and V. Sharko.

Similar Morse theory of many-valued functions and inequalities is constructed (see [60]) also for the case where the degree of irrationality of the form ω is greater than zero, i.e., when $k > 1$ (see above). The general construction scheme is the same, though there appear some new interesting effects related to rings of Laurent series in many variables.

We now discuss another problem which is closely connected with many-valued mappings characterizing some functionals [27]. In Chapter IV. we are going to consider two-dimensional minimal surfaces in the Euclidean space $\mathbf{R}^3$, which span closed contours homeomorphic to the circle. Under a contour deformation, the surface is changed; in particular, its area is changed. Thus, we can consider an area graph of a minimal surface defined on the space of all boundary contours in $\mathbf{R}^3$. Since, generally speaking, several minimal surfaces can span the same contour, the area graph turns out to be a many-valued function of the contour. Now, we formulate the general problem. Consider the space $C^\infty(M, N)$ of smooth mappings $f : M \to N$, where N is a fixed Riemannian manifold with non-vanishing boundary ∂M. On the space $C^\infty(M, N)$, consider a functional $F(f)$ given by $F(f) = \int_M L(f) d\sigma$, where L is the Lagrangian depending on the mapping f and its derivatives, $d\sigma$ is the Riemannian volume form on the manifold M. As an example, consider the volume (or area) functional vol of the image $f(M)$ in N. Consider the Euler-Lagrange equation $I(f) = 0$ for the functional F. Extremals f_0 of the functional F are the solution of the Euler-Lagrange equation $I(f) = 0$. For example, locally minimal surfaces are extremals of the volume functional. Two-dimensional minimal surfaces are modeled by soap films in $\mathbf{R}^3$, spanning closed wire contours. To specify an extremal f_0, it is necessary to give boundary conditions, i.e., the mapping $\gamma : \partial M \to N$. For example, to specify a soap film, it is necessary to fix in $\mathbf{R}^3$ a closed contour γ, i.e. to fix a mapping of the circle S^1 in $\mathbf{R}^3$. The set of all boundary conditions is identified naturally with the space $C^\infty(\partial M, N)$ of mappings $\gamma : \partial M \to N$. A boundary condition, generally speaking, determines some extremals f_0 of the functional F, i.e., several solutions of the Euler-Lagrange equations, so that $f_0|_{\partial M} = \gamma$. Denote the set of all extremals with given boundary condition γ by K_γ. Computing the value of the functional F for each extremal, we obtain the set of numbers $\{F(f_0)\}$. As a result, we get a many-valued mapping (function) $\gamma \to \{F(f_0)\}$, where $f_0 \in K_\gamma$. This function is defined on the space $C^\infty(\partial M, N)$ having, generally speaking, branch points and other singularities. Various functionals (and their Euler-Lagrange equations) are characterized, generally speaking, by different types of singularities. The general problem is what singularities characterize some or other Euler-Lagrange equations. While making a concrete study of many-valued functions of the form $\gamma \to \{F(f_0)\}$, it is useful at first to simplify the problem by considering finite-dimensional subspaces in $C^\infty(\partial M, N)$ on which the function $\gamma \to \{F(f_0)\}$ should be bounded, instead of the whole space $C^\infty(\partial M, N)$ for the area functional of two-dimensional soap films. A simpler version of the general problem is that of the length (or action) functional defined on the space of piecewise paths joining together a pair of points on a smooth connected Riemannian manifold. As a result, we obtain a many-valued function on $N \times N$, where N is a given manifold. Two sheets of this many-valued function intersect if there are two geodesics of the same length joining together a pair of points (geometry of the geodesics is different). Here, the extremals of the functional are geodesics with fixed ends. This many-valued function has an interesting structure even for the simple case of a flat torus.

CHAPTER III/MANIFOLDS OF SMALL DIMENSIONS

1. Homeomorphisms of Two-Dimensional surfaces

1. Homeotopy Group and Classification of Three-Dimensional Manifolds of Genus 1

In this chapter, we shall be guided to some extent by the material given in [34], for example. Denote by Π a three-dimensional manifold with boundary, bounded in $\mathbf{R}^3$ by a standardly embedded two-dimensional sphere M_g^2 with g handles (Fig. 1). In the diagram, $g = 3$. For brevity, the manifold Π is said to be a complete pretzel of genus g. Consider two copies of complete pretzel Π_1 and Π_2. It is well known that if we consider an arbitrary homeomorphism $\alpha : \partial \Pi_1 \rightarrow \partial \Pi_2$, we can construct a three-dimensional closed compact connected and orientable manifold $M(\alpha)$. It suffices to paste together two complete pretzels Π_1 and Π_2 along the homeomorphism of their boundaries. The converse is also valid: Any three-dimensional manifold of given type can be represented as $M(\alpha)$ for a certain homeomorphism $\alpha : \partial \Pi_1 \rightarrow \partial \Pi_2$ of the surfaces of genus g. Such a representation is non-unique.

Among the set of all homeomorphisms $\alpha : \partial \Pi_1 \rightarrow \partial \Pi_2$, we distinguish homeomorphisms which can be extended to the interior, i.e., to a homeomorphism of complete pretzels $\hat{\alpha} : \Pi_1 \rightarrow \Pi_2$.

LEMMA 1. *If $\alpha_1, \alpha_2 : \partial \Pi_1 \rightarrow \partial \Pi_2$ are homeomorphisms extendible to the interior of a sphere with g handles, then three-dimensional manifolds $M(\alpha)$ and $M(\alpha_2 \alpha \alpha_1)$ are homeomorphic for any homeomorphism $\alpha : \partial \Pi \rightarrow \partial \Pi$.*

PROOF. Define a continuous mapping $\varphi : M(\alpha_2 \alpha \alpha_1) \rightarrow M(\alpha)$ in the following way:

$$\varphi(x) = \begin{cases} \hat{\alpha}_1(x) & \text{if } x \in \Pi_1 \\ \hat{\alpha}_2(x) & \text{if } x \in \Pi_2 \end{cases}.$$

Here, we denote the extensions of homeomorphisms α_i by $\hat{\alpha}_i$. It is easy to verify that the mapping φ is a homeomorphism.

$$M_g^2 = \partial H$$

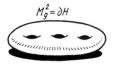

Figure 1

Recall that two homeomorphisms $\alpha_1, \alpha_2 : \partial\Pi \to \partial\Pi$ are said to be isotopic if they are homotopic and if there exists a homotopy f_t connecting α_1 and α_2, i.e., $f_0 = \alpha_1, f_1 = \alpha_2$, and such that each mapping $f_t : \partial\Pi \to \partial\Pi$ is a homeomorphism.

LEMMA 2. *If the homeomorphism* $\alpha : \partial\Pi \to \partial\Pi$ *is isotopic to the identity mapping of* $\partial\Pi$ *in itself, then* α *is extendible to the interior, i.e., to a homeomorphism of a complete pretzel.*

The proof follows from the fact that the identity mapping of the boundary of a complete pretzel in itself is probably extendible to the interior.

Consider a group of all homeomorphisms of a two-dimensional surface $\partial\Pi$ of genus g. Denote this group by H_g. In this group, distinguish a subgroup I_g of homeomorphisms which are isotopic to the identity homeomorphism.

LEMMA 3. *The group* I_g *composed of homeomorphisms of two-dimensional surface* $\partial\Pi$, *isotopic to the identity homeomorphism, is an invariant subgroup in the group of all homeomorphisms* H_g.

This statement is obvious. Therefore, the quotient group $\Gamma_g = H_g/I_g$ called the *homeotopy group* of the two-dimensional pretzel $\partial\Pi$ of genus g is defined correctly. Compute this group for the case of $g = 1$. Construct a homomorphism ψ of the homeotopy group Γ_1 into the group of all automorphisms of the Abelian group $Z \oplus Z$ realized as the fundamental group of a torus T^2, i.e., $Z \oplus Z = \pi_1(S^1 \times S^1)$. The above group of automorphisms coincides with the group of invertible integral matrices of order two with unit discriminant, which we denote by $SL(2, Z)$. The homomorphism ψ is defined in the following way: Each homeomorphism α of the surface induces an automorphism $\psi(\alpha)$ of its fundamental group $Z \oplus Z$. It is clear that a homeomorphism isotopic to the identity induces the identity automorphism.

THEOREM 1. *A homomorphism* $\psi : \Gamma_1 \to SL(2, Z)$ *is an isomorphism.*

PROOF. The following matrices are generating in the group $SL(2, Z)$:

$$\begin{pmatrix} \pm 1 & 0 \\ 0 & \pm 1 \end{pmatrix}, \begin{pmatrix} 1 & 1 \\ 0 & 1 \end{pmatrix}, \begin{pmatrix} 1 & 0 \\ 1 & 1 \end{pmatrix}.$$

To prove the surjectivity of a mapping ψ, it suffices to represent these matrices as the automorphisms of the integral lattice $Z \oplus Z$, generated by homeomorphisms of a torus. It is clear that the following mappings of the torus onto itself should be taken as such homeomorphisms: (a) a reflection of the torus with respect to the yz-plane (Fig. 2(a)); (b) a reflection of the torus with respect to the xy-plane (Fig. 2(b)); (c) a transformation of the torus, i.e., twisting round the angle 2π shown in Fig. 2(c) and called the Dehn operation; (d) a transformation shown in Fig. 2(d) (also, the Dehn transformation). Clarify (c) and (d). In view of their similarity, it suffices to comment on one operation, e.g., (c). Consider the torus meridian (vertical section), cut the torus

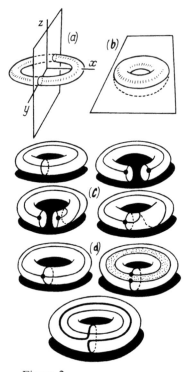

Figure 2

along this curve, then twist the right part of the obtained cylinder round the angle 2π, and paste it again to the left boundary of the cylinder. It is clear that the parallel which had type $(1,0)$ will transform to a trajectory of type $(1,1)$ traversing the torus once along the parallel and once along the meridian. Let us prove the monomorphity of the mapping ψ.

Given a homeomorphism of the torus such that the induced automorphism of the fundamental group is the identity mapping. We must prove that the homeomorphism is isotopic to the identity.

LEMMA 4. *If two closed tame non-self-intersecting curves on a torus do not decompose the torus (into the union of two non-intersecting domains) and if they are homotopic, they are isotopic.*

PROOF. Let γ_1 and γ_2 be two such curves. Since they are homotopic, their intersection number is zero. (See, e.g., the definition and intersection number properties in [1, 4, 34]). First, consider the case where the curves intersect. At each point, a sign of intersection is defined. By adding up these ± 1, we obtain a zero intersection number due to the ascribed homotopy. Let A and B be two neighboring points of opposite signs, which are the intersection points of γ_1 and γ_2. Let l be an arc connecting these two neighboring points (Fig. 3). Cut

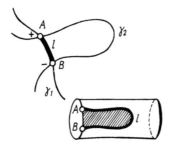

Figure 3

the torus along the curve γ_2. Since this curve does not decompose the torus, the cut results in a cylinder with the points A and B being on one of the boundary circles. As the arc l bounds part of the cylinder, homeomorphic to the disk, we can "remove" the arc l by means of an isotopy, eliminating two intersection points of opposite signs. Thus, we have decreased a number of intersection points of the curves. Carrying on this process, we eliminate all intersection points of both curves. Therefore, using the corresponding isotopy, we can assume that the curves γ_1 and γ_2 do not intersect. Since they do not decompose the torus, we cut it along one of the curves, e.g., γ_2, and obtain a cylinder with the curve γ_1 winding around it without self-intersections. It is clear that there exists an isotopy making γ_1 coincident with γ_2. Q.E.D.

Let us return to the proof of the theorem. Consider two standard generators of the group $\pi_1(T^2)$: a parallel b and meridian a. Fix the meridian a, for example and consider its image under a homeomorphism α. The curve $\alpha(a)$ and the curve a belong to the same homotopy class, (given); hence, they are homotopic. Obviously, both curves do not decompose the torus. Therefore, by Lemma 4, there exists an isotopy carrying $\alpha(a)$ back into a. Similarly, we prove the existence of an isotopy carrying a curve $\alpha(b)$ back into its position b. One should be careful not to displace a. Thus, carrying the meridian and then the parallel into the original position, we obtain the identity torus homeomorphism on the generators a and b, i.e., on the one-dimensional skeleton of the torus. Since the complement to the union of these curves on the torus is a disk, it is not difficult to complete the construction of the required isotopy on a two-dimensional disk, the homeomorphism on the disk boundary having been transformed into the identity by means of an isotopy. The theorem is proved.

Thus, if Q_1 is the set of all three-dimensional manifolds representable in the form of $M(\alpha)$, where α is a homeomorphism of the torus onto itself (i.e., we consider pretzels of genus 1), an epimorphism $B(2, Z)\backslash SL(2, Z)/B(2, Z) \to Q_1$. Here, by $B(2, Z)$ we denote the matrix subgroup of the group $SL(2, Z)$, which possess the following property: Homeomorphisms of the torus, corresponding to the matrices, are extendible to the interior. The symbol with two oblique lines denotes the set of cosets generated by right and left translations through the elements of the subgroup $B(2, Z)$. Here, we are guided by Lemma 1.

LEMMA 5. *The matrix* $\begin{pmatrix} a & b \\ c & d \end{pmatrix} \in SL(2, Z)$ *belongs to the subgroup* $B(2, Z)$ *if and only if* $c = 0$.

PROOF. The necessity follows from the fact that the image of the meridian must contract in the solid torus, i.e., it must not wind round the parallel. The sufficiency follows from a simple observation that a disk can span the meridian image, then the required homeomorphism is extended via radii to the remaining three-dimensional cell.

Consider the operation of integral matrix right and left multiplication by matrices of the form $\begin{pmatrix} \pm 1 & \pm k \\ 0 & \pm 1 \end{pmatrix}$, i.e., by the elements of the subgroup $B(2, Z)$. Recall that $\det \begin{pmatrix} a & b \\ c & d \end{pmatrix} = \pm 1$. It is clear that this operation is equivalent to the change in sign ascribed to rows and columns and to adding the first column to the second or the second row to the first one (with coefficient k).

LEMMA 6. *Any matrix from the group* $SL(2, Z)$ *is reduced by the above left and right multiplication by the matrices from* $B(2, Z)$ *to one of the following matrices*:

(1) $\begin{pmatrix} 1 & 0 \\ 0 & 1 \end{pmatrix}$;   (2) $\begin{pmatrix} 0 & 1 \\ 1 & 0 \end{pmatrix}$;   (3) $\begin{pmatrix} q & x \\ p & y \end{pmatrix}$, *where* $p \geqslant 2$, $0 < q < p$, $qy - px = 1$.

PROOF. (1) If the original matrix is such that there is a zero in its lower left corner, then it has the form $\begin{pmatrix} \pm 1 & \pm k \\ 0 & \pm 1 \end{pmatrix}$ and is apparently reduced by the above operations to the form $\begin{pmatrix} 1 & 0 \\ 0 & 1 \end{pmatrix}$. (2) If there is 1 in the lower left corner, first, we should obtain zeros on the matrix diagonal (by means of the above operations). In this case, there automatically appears ± 1 in the right upper corner. (3) Matrices which do not satisfy the conditions of points (1) and (2) are obviously reduced to (3).

It is easily checked (we leave the verification to the reader) that there are the following homeomorphisms: if $\psi(\alpha) = \begin{pmatrix} 1 & 0 \\ 0 & 1 \end{pmatrix}$, then $M(\alpha) = S^2 \times S^1$; if $\psi(\alpha) = \begin{pmatrix} 0 & 1 \\ 1 & 0 \end{pmatrix}$, then $M(\alpha) = S^3$.

If $\psi(\alpha) = \begin{pmatrix} q & x \\ p & y \end{pmatrix}$ is a matrix of form (3), the corresponding manifold $M(\alpha)$ is denoted by $L(p, q)$, and called a *lens*.

THEOREM 2. *Lenses $L(p,q)$ and $L(p_1,q_1)$ are homeomorphic if and only if the relations* (1) $p=p_1$ *and* (2) $qq_1 \equiv \pm 1 \pmod{p}$ *or* $q \equiv \pm q_1 \pmod{p}$ *hold.*

In one direction, this theorem is proved easily. Let us prove that if conditions (1) and (2) are fulfilled, then the lenses $L(p,q)$ and $L(p_1,q_1)$ are homeomorphic. Consider the case where $p=p_1$ and $q=\pm q_1 \pmod{p}$, i.e., $q=\pm q_1 + mp$ for an integer m. Then the matrix $\begin{pmatrix} \pm q \pm mp & x_1 \\ p & y_1 \end{pmatrix}$ is obviously obtained from the matrix $\begin{pmatrix} q & x \\ p & y \end{pmatrix}$ by means of the above operations. Since these operations do not change the type of a manifold, the corresponding lenses are homeomorphic. If $p=p_1$ and $qq_1 = \pm 1 + mp$, then, by replacing the matrix $\begin{pmatrix} q & x \\ p & y \end{pmatrix}$ with the inverse one, we do not change the corresponding manifold since it is clear that $M(\alpha)=M(\alpha^{-1})$, while the inverse matrix has the form $\begin{pmatrix} y & -p \\ -x & q \end{pmatrix}$, i.e., q has been replaced by its "inverse" element y, for $qy=1+px$. The proof of the converse statement of the theorem is less trivial, and we do not give it here.

2. A System of Homotopy Group Generators of the Solid Pretzel (Handlebody)

As we have seen already, discussing the homeotopy group Γ_g of a two-dimensional pretzel $\partial \Pi$ of genus g enables us to study three-dimensional manifolds admitting representation of the form $M(\alpha)$, where α is a homeomorphism of the surface $\partial \Pi$ of genus g onto itself. To be more precise, let $K_g \subset H_g$ be the subgroup of the group H_g of homeomorphisms of the two-dimensional surface $\partial \Pi$ composed of homeomorphisms of the two-dimensional pretzel $\partial \Pi$ of genus g, extendible to the solid pretzel Π. Let I'_g be the group of homeomorphisms of a solid pretzel, which are isotopic to the identity. By Lemma 2, we obtain an inclusion $I_g \subset K_g$; therefore, the group $G_g = K_g/I'_g$ is a subgroup of $\Gamma_g = H_g/I_g$. The group G_g may be interpreted as the solid pretzel homeotopy group. By Lemma 1, we obtain the following corollary: There exists the mapping

$$G_g \backslash \Gamma_g / G_g \to Q_g,$$

which is "onto", where by Q_g we denote the set of three-dimensional manifolds of form $M(\alpha)$, where α is a homeomorphism of genus g and $G_g \backslash \Gamma_g / G_g$ is the set of double cosets, i.e., generated by homeomorphisms of the form $\alpha_1 h \alpha_2$, $(\alpha_i \in G_g, \ h \in \Gamma_g)$. From the standpoint of describing the set of all three-dimensional manifolds, a description of the group Γ_g and its subgroup

G_g is, therefore, of considerable interest. To use the introduced notation conveniently, we give the following table:

Solid pretzel Π	Its boundary $\partial\Pi$	Group of
K'_g	H_g	homeomorphisms
I'_g	I_g	homeomorphisms isotopic to identity
–	K_g	homeomorphisms extendible to the interior
$G_g = K_g/I'_g$	$\Gamma_g = H_g/I_g$	homeotopies

It turns out that the *Dehn twistings* around the curves s_i, 1_i, $1 \leqslant i \leqslant g$ and m_i, $1 \leqslant i \leqslant g-1$ ($3g-1$ curves in all, Fig. 4) and symmetries with respect to plane sections have generated the group Γ_g.

Denote by p a homeomorphism of the pretzel $\partial\Pi$, fixed on the curves s_i, $2 \leqslant i \leqslant g$ and carrying the curve s_1 into the curve m_1. It is clear that such a homeomorphism exists and extends to the interior. Consider the group S_g of homeomorphisms of a complete pretzel, which possess the property of carrying the set of curves s_i, $1 \leqslant i \leqslant g$ into itself, i.e., they are invariant on the set of the curves s_i. It is clear that $S_g \subset K_g$. $S_g \subset K_g$. The image of this group under the projection $K_g \to G_g$ is denoted by $[S_g]$.

THEOREM 3. *The group G_g is generated by all elements of the group $[S_g]$ and an element $[p]$. In addition, the group G_g has a finite number of generators.*

This theorem was proved in 1973 by S. V. Matveev. The proof can be found in [35].

It is clear that a more detailed analysis enables us to simplify this description. Represent a solid pretzel as shown in Fig. 5. Let a_1 be a homeomorphism of this pretzel, which is a rotation of the pretzel through an angle $2\pi/g$ around the x-axis. By a_2, a_3, a_4, denote homeomorphisms of the solid pretzel, resulting from the following operations: (a) In the case of a_2 and

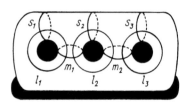

Figure 4

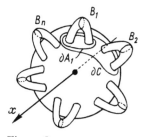

Figure 5

a_3, twist the pretzel through an angle π around the circles ∂A_1 and ∂C, respectively; (b) In the case of a_4, twist the pretzel through an angle 2π around the circle ∂B_1. Then put $a_5 = p$, where the homeomorphism p has been defined above. Finally, as a_6, take any homeomorphism of the solid pretzel, with a change in orientation.

THEOREM 4. *A homeotopy group of a solid pretzel is generated by six elements* $a_1, a_2, \ldots, a_6$. *A group of orientation-preserving homeotopy of a solid pretzel is generated by the five elements* $a_1, a_2, \ldots, a_5$.

2. An Algorithm of Recognizing the Standard Three-Dimensional Sphere in the Class of Heegard Diagrams of Genus Two

1. The Whitehead Graphs Encoding Three-Dimensional Manifolds. Operations of Indices One and Two

As has been already stated, any three-dimensional closed manifold may be represented (non-uniquely) in the form $M(\alpha)$, i.e., in the form of gluing two solid pretzels Π_1, Π_2 of genus g together with respect to a diffeomorphism of their boundaries $\alpha : \partial\Pi_1 \to \partial\Pi_2$. A representation of a manifold in such a form is sometimes called the *Heegard diagram* of a three-dimensional manifold. Fix on a two-dimensional pretzel the set of standard meridians $a_1, \ldots, a_g$ (Fig. 6). Consider their images under the diffeomorphism α, and denote them by

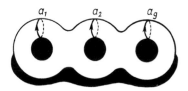

Figure 6

$\bar{a}_1, \ldots, a_g, \bar{a}_i = \alpha(a_i)$. Without loss of generality, the circles a_i and $\bar{a}_j$ can be regarded as intersecting transversally; in particular, in a finite number of points.

DEFINITION 1 We will say that a system of $2g$ circles $a_1, \ldots, a_g, \bar{a}_1, \ldots, \bar{a}_g$ generates a net s on a two-dimensional pretzel. The number g is called the genus of the net s. We refer to the circles a_i as the *circles of index one*, and $\bar{a}_j$ as the *circles of index two*.

This terminology is due to the relation of the nets to Morse functions on three-dimensional manifolds. We do not need the analysis here. The interested reader is referred to [40]. By the definition of the net, it follows that a closed manifold defined up to a homeomorphism corresponds to each net. In addition, each three-dimensional manifold M determines an infinite set of nets corresponding to all diffeomorphisms α such that $M = M(\alpha)$. In this sense, we can regard the set of nets as the set of codes which are used for encoding three-dimensional manifolds. Note the property of the net which will be of use in the sequel. Cut the pretzel along all circles of index one. It is evident that we obtain a two-dimensional sphere with $2g$ holes. Each of the circles a_i generates two edges of the cut, denoted by a_i and a_i' (with orientation preserved). Similarly, cutting the pretzel along all circles of index two, we obtain a sphere with $2g$ holes. Let us assume that, on the circles of index one, the standard orientation was fixed (Fig. 6); then all circles of index two are oriented, too. In cutting the pretzel along all circles of index one, the circles of index two break apart, and transform into a set of oriented line segments joining the points together on the boundaries. Supply each line segment with orientation and the number specified by a circle of index two, of which this line segment is a part. Hereby, it turns out that, on a two-dimensional sphere with $2g$ holes, we obtain the set of smooth arcs joining the points on the hole boundaries together. The example in Fig. 7 shows a pretzel of genus three and two graphs obtained by cutting along the circles of index one (Fig. 7(c)) and along the circles of index two (Fig. 7(e)). For convenient representation, the sphere with $2g$ holes, a point not belonging to the graphs has been cut out, resulting in a plane disk with $2g$ holes. In the sequel, we show the obtained graphs on the orginary plane from which $2g$ non-intersecting disks are cut out.

DEFINITION 2 Let α be an arbitrary diffeomorphism of pretzel of genus g. Consider the net s corresponding to it. A *Whitehead graph* $W_i(s)$ of index i is the graph obtained on the plane by means of cutting the pretzel along the circles of index i. Thus, each diffeomorphism α determines two Whitehead graphs $W_1(s)$ and $W_2(s)$, where $s = s(\alpha)$.

Generally speaking, these two graphs do not coincide. The vertices of a Whitehead graph are the cut circles, the edges are parts of circles of the other index, which are broken by the cuts into a set of segments. We assume that all

the beginning of this century, Poincaré raised a question how to recognize the standard sphere in a list of all three-dimensional manifolds (see, e.g., the details in [34]). Closely connected with this problem is the famous Poincaré problem: Is it true that any three-dimensional homotopy sphere is homeomorphic to the standard one? Recall that a manifold is called a homotopy sphere if it has the same homotopy groups as the sphere.

In this section, we concentrate on a problem of algorithmic recognition of a three-dimensional sphere in the class of Heegard diagrams of genus two. In other words, it is required to find an algorithmically recognizable (and easily computed) characteristic property of Whitehead graphs which correspond to the standard sphere. If such a property is found, then the following question will be answered: Do an arbitrary given pair of Whitehead graphs correspond to the standard sphere, or do they describe another manifold which is non-homeomorphic to the sphere? We do not analyze the subtlety of algorithmic recognition here, since we confine ourselves to intuitive representation, according to which we have to "design' a computer with a general program answering the question if the code of a manifold, supplied to the computer "input", is the code of the standard sphere or not (see [40–45] for different approaches to solving this problem).

It turn out that, in some important particular cases, the required algorithm of recognizing the standard sphere exists. In 1973, Birman and Hilden in their fine work [70] reduced the problem of recognizing the sphere S^3 in the class of Heegard diagrams of genus two to a well-known *Haken algorithm* of recognizing the trivial knot. Thus, Birman and Hilden proved in principle the algorithmic solubility of the problem of recognizing a sphere in the above class of Heegard diagrams. In fact, they proved a stronger statement: A class of diagrams, in which the codes of the standard sphere are recognized theoretically, is somewhat wider than the class of Heegard diagrams of genus two, though the Haken theoretical algorithm of recognizing the trivial knot is rather abstract and complex; e.g., to realize it by a computer is, probably, not yet possible. Therefore, the problem of recognizing the standard sphere effectively by a simple and visual algorithm (admitting a simple computer realization) still remained. Finally, such an algorithm was found by A. T. Fomenko and I. A. Volodin in 1973–1974; later, A. T. Fomenko, V. E. Kuznetsov, and I. A. Volodin realized it on a *BESM-6* computer. In 1977, a new extended series of computer experiments was made by A. T. Fomenko. The algorithm was found as a result of a correct "reencoding" of three-dimensional manifolds and discovering an important recognizable topological "characteristic property", i.e., a certain constructive invariant of standard sphere codes. A new method of encoding manifolds is, after all, equivalent to the encoding by means of classical Heegard diagrams, though it made explicit the characteristic property of standard sphere codes which admit an effective recognition in the class of Heegard diagrams of genus two.

For the use in the sequel, it is convenient to consider a *net* up to an arbitrary diffeomorphism of the pretzel onto itself. In other words, the circles a_i need not be standard meridians now; a_i can be the images of the latter under an arbitrary diffeomorphism. If two nets are equivalent in this sense, i.e., carried into each other by a diffeomorphism, it is easy to verify that the corresponding manifolds are diffeomorphic. Therefore, replacing the net with its equivalent, we do not change the topological types of the manifold; instead, it is not now necessary to recall always that a_i are represented by standard meridians. In particular, we remove a certain inequality between circles of index one and two, arising at the moment of using the circles a_i as the basis for the net definition. In case of need, we can always make circles of index one transform into standard meridians by means of applying a convenient diffeomorphism. Similarly, circles of index two can be transformed into standard parallels. Naturally, circles of index one are not mapped into standard meridians, but transform to some, possibly, complex circles on the pretzel.

Here is an example of the simplest net s determining the standard sphere. Let us take standard meridians as a_i, and standard parallels as $\bar{a}_j$, A diffeomorphism of the pretzel, giving such a net (not determined uniquely) interchanges with parallels and meridians. Pasting two solid pretzels together with respect to such a diffeomorphism and using the information in Sec. 1 of the present chapter, we obtain the standard sphere. Figure 8 shows Whitehead graphs corresponding to this net. They are the simplest in the set of all Whitehead graphs. It turns out that any net s of the standard sphere can be obtained from the simplest net s_0 (of the same genus) by applying to the net s_0 some simple operations which we are going to describe.

Let s be a net on a manifold M, and not on a sphere. On the pretzel, consider two circles of index one, a_i and a_j. Let γ be a smooth path connecting a point of the circle a_j with a point of the circle a_i and having no other intersection points with circles of index one. Replace a pair of the circles a_i, a_j with a new pair of

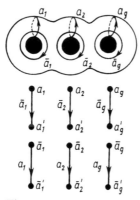

Figure 8

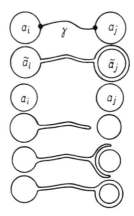

Figure 9

the circles $\tilde{a}_i, \tilde{a}_j$, where $\tilde{a}_j = a_j$. A new circle $\tilde{a}_i$ is shown in Fig. 9. In other words, from the circle a_i, a thin tongue grows, slides along the path γ, approaches the circle a_j, where it forks, and begins encircling a_j. Finally, it "swallows" the circle a_j. The final result is shown in Fig. 9. The circle a_j has not been changed.

DEFINITION 3 The operation described above is referred to as an *operation of index one*. Similarly, an *operation of index two* is defined.

Note that the segment γ from the definition of an operation of index one could intersect the circles of index two. As always, we consider all intersections to be transversal, i.e., being in a general position.

LEMMA 1. *Let some operations of indices one and two in an arbitrary order be applied to a net s determining a manifold M. As a result, we obtain a net (in the sense of Definition 1) of the same genus as the original net. Furthermore, this new net determines a manifold diffeomorphic to the original one. Thus, the operations of indices one and two do not change the topological type of a three-dimensional manifold.*

This lemma can be proved by the methods of Morse theory. (See, e.g., the proof in [40]).

Thus, we can, proceeding from a fixed net, construct an infinite set of nets giving the same manifold, whereas the operations of indices one and two can be used in arbitrary combinations. These operations in the general case do not enable us to obtain all nets of the given manifold (and of the given genus) proceeding from one net. But, in one case, it works (Waldhausen).

PROPOSITION 1. Let s_0 be an ordinary net of genus g of the standard sphere shown in Fig. 8. Then any other net of the standard sphere of the same genus may be obtained from the net s_0 as a result of applying some operations of indices one and two.

See, e.g., the proof in [43, 40]. Thus, a set of nets obtained from an ordinary net by applying the operations of indices one and two coincides exactly with the set of all nets corresponding to the standard sphere. An important fact is that any net of a fixed genus can be obtained by the operations of the described type from an ordinary net of the same genus. Note that, applying the operations of indices one and two, we do not change the genus of the net. Proposition 1 describes all codes of the standard sphere in the space of all codes of all three-dimensional manifolds. Nevertheless, one should not think that it gives an algorithm of recognizing the codes of a sphere. The thing is that, in addition to the aforesaid, we have to learn to recognize the nets obtained from the ordinary one by applying the described operations.

2. Waves Separating the Vertices and the Operation of Reducing Whitehead Graphs

Consider a net on a pretzel of some genus. Recall that we consider the oriented nets only (point 1). The net determines the pretzel decomposition into some domains U.

DEFINITION 4 A domain U is said to be *distinguished* if, among the edges determining its boundary, there are two edges β_1 and β_2 belonging to the same circle, and if their orientations are compatible with one of the traverses of the domain boundary. The edges β_1 and β_2 are referred to as distinguished (Fig. 10). A line segment inside the distinguished domain, connecting interior points of the distinguished edges β_1 and β_2 is called a *wave* τ (Fig. 10).

The notion of a wave was introduced by A. T. Fomenko and I. A. Volodin. Its properties are studied in detail in [40]. This notion has turned out to be useful for constructin an algorithm for recognizing Heegard diagrams of a sphere (for genus two). Let us reformulate the definition of the wave in the

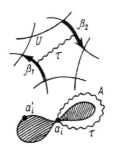

Figure 10

language of Whitehead graphs. It appears that the presence of the wave in the net is particularly obvious in these graphs.

For definiteness, let the distinguished edges (see Definition 4) belong to the circle a_i of index one. Cut the pretzel along the circle of index one. As a result, we obtain a graph $W_1(s)$. The circle a_i transforms into two vertices a_i and a_i' of the graph $W_1(s)$. The wave, i.e., the line segment τ, transforms into a smooth line segment beginning and ending at one of these vertices; in which one exactly, depends on the relationship between the orientations. For definiteness, let τ begin and end at a_i. A remarkable property of the arc τ is that it intersects the graph $W_1(s)$ at the point a_i only (Fig. 10). Therefore, removing the vertex a_i from the graph, we separate it into at least two connected nonvoid components. One of them is in the domain A bounded on the plane by the wave τ; another is outside this domain (Fig. 10). We assume that $a_i' \notin A$ and a_i lies at least on one of the edges from E.

Definition 5. The vertex of a Whitehead graph, in which the wave τ begins and ends, is called a *separating vertex* if there exists at least one edge of the graph, which is in A and contains the vertex a_i.

Such vertices appeared first in the fine work by Whitehead [42], in which he solved the problem of recognizing the sets of elements in a free group, which are its generators (the so-called simple sets).

The role of a separating vertex lies in the fact that its presence on one of Whitehead graphs enables us to simplify both Whitehead graphs and, therefore, the Heegard diagram of a manifold. Equivalently, the presence of a wave in the net enables us to reduce the net and, by means of operations of indices one and two, pass on to a new net with fewer edges. Now, we describe the procedure of reducing a Whitehead graph.

Let a_i be a separating vertex on the graph $W_1(s)$. The wave τ separates the plane into two domains A and B, one of which contains a vertex a_i'. Recall that each circle a_i generates two vertices a_i and a_i' when the pretzel is cut.

Let A be a domain which does not contain a_i'. According to the definition of a separating vertex, there exists in A at least one edge of the graph which contains the vertex a_i. We say that an edge of the Whitehead graph is incident with a vertex if the edge belongs to the vertex or goes out of it. Consider the set of all edges of the Whitehead graph, incident with the vertex a_i and situated in the domain A enclosed by the wave (Fig. 11). These edges are grouped in three classes.

(1) Edges of the first class are parts of the circle (of index two), interacting with a_i in the following way: This part ($\bar{a}$ in Fig. 11) passes through a_i, "jumps over" to the vertex a_i' without passing through any other graph vertices. Further, returning to a_i', $\bar{a}$ goes back to a_i and, emanating from it, returns to the part enclosed by the wave.

(2) Edges of the second class are parts of the circle which interacts with a_i as follows: This part ($\bar{b}$ in Fig. 11) enters a_i, "jumps over" to the vertex a_i', then reaches some other vertex, different form a_i' and outside the domain A.

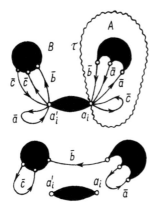

Figure 11

(3) Edges of the third class are parts of the circle which interacts with a_i in the following way: This part ($\bar{c}$ in Fig. 11) emanates from some vertex situated outside A, goes to a_i', "jumps over" to the vertex a_i, makes a loop, and returns to a_i again without passing through any other vertices. Part $\bar{c}$ "jumps over" to a_i and, emanating, passes through some vertex outside A.

We describe the *surgery*, the operation of simplifying the Whitehead graph (at least one separating vertex a_i is given). This operation transforms the Whitehead graph in the following way: (1) In the case of the edges of the first class, loop $\bar{a}$ beginning and ending in the vertex a_i' is annihilated; two edges incident with the vertex a_i are annihilated and replaced by an edge joining the vertices of the graph, different from a_i, with which these edges were originally incident (Fig. 11). (2) In the case of the edges of the second class, both edges $\bar{b}$ incident with the vertices a_i and a_i', respectively, are annihilated and replaced by an edge which, leaving out the vertices a_i and a_i', joins directly the vertices of the graph, different from a_i and a_i', with which these edges were originally incident (Fig. 11). (3) In the case of the edges of the third class, the operation coincides with that of the edges of the first class. The only distinction is that the roles of the vertices a_i and a_i' should be exchanged (Fig. 11).

Though the Whitehead graph surgery has been fully described, it needs some comments, since the description is abstract, e.g., without regard to the a priori possibility of realizing the above surgery being not at all obvious.

THEOREM 1 (due to A. T. Fomenko and I. A. Volodin. See [40]). (*a*) *Let $\tilde{W}$ be a graph obtained from the Whitehead graph $W_i(s)$ by means of the above surgery in the presence of at least one separating vertex. Then, the graph $\tilde{W}$ is a plane graph, i.e., it admits a realization on the plane, and moreover, all above surgeries could be realized in the plane from the very beginning.* (*b*) *The graph $\tilde{W}$ is a Whitehead graph for a certain new net $\tilde{s}$, i.e. $\tilde{W} = W_i(\tilde{s})$.* (*c*) *The new net $\tilde{s}$ is obtained from the original net s (to which corresponded the original Whitehead* *graph) by operations of indices one and two (in fact, operations of one index are sufficient). (d) The net $\tilde{s}$ determines the same manifold (up to diffeomorphism) as the original net s, i.e., the above surgery preserves the manifold topological type. (e) The new Whitehead graph $W_i(\tilde{s})$ has fewer edges than the original graph (with the same number of vertices); therefore, it is simpler.*

By (e) of Theorem 1, the surgery operation is called the *Whitehead graph simplification operation*. The new graph $W_i(\tilde{s})$ is simpler than the original in the sense that the number of its edges is strictly fewer than the number of edges of the original graph. Statement (e) is evident from Fig. 11 which shows that the surgery reduces the number of edges at least by one. Statement (d) obviously follows from statement (c) and Lemma 1. Statements (b) and (c) are more complex (see [40]). Thus, if the net s of a manifold M^3 contains a wave (i.e., if there is a separating vertex on at least one of its Whitehead graphs $W_i(s)$), then an extremely simple algorithm enables us to carry the net s into a new net $\tilde{s}$ which determines the same manifold and is simpler than the original net.

3. The Existence of a Separating Vertex on the Whitehead Graphs of Genus Two, Corresponding to the Standard Three- Dimensional Sphere. An Algorithm of Recognizing the Sphere

The simplest net s_0 corresponding to the standard sphere does not contain a single separating vertex (Fig. 8). However, the net in Fig. 7 corresponds to the standard sphere (verify!), and its both Whitehead graphs contain separating vertices (Fig. 8). Moreover, we can prove that if the net s is obtained from the simplest net s_0 by operations of the same index, then s must contain at least one wave (under the condition that $s \neq s_0$). Meanwhile, we should bear in mind that the separating vertex can be on only one of two Whitehead graphs. A simple example of the net corresponding to the standard sphere is shown in Fig. 12. One of the two Whitehead graphs contains two separating vertices; the other, none (Fig. 12(b)).

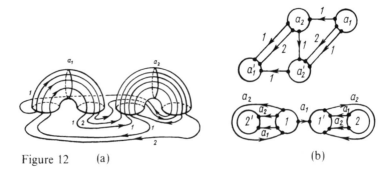

Figure 12   (a)   (b)

As a result of the analysis of the fundamental work by Whitehead [42], I. A. Volodin and A. T. Fomenko formulated the following hypothesis: *Any net corresponding to the standard sphere and different from the simplest net contains at least one wave* [40], [48]. To verify this hypothesis, a *BESM*-computer was used which simulated the process of constructing nets by operations of indices one and two [40]. It turned out that all 10^6 nets constructed by the computer at random and giving the standard sphere do contain at least one "wave". Then it was found theoretically (see [47]) that this hypothesis is nevertheless incorrect for the nets of genus more than two (see also [49], [50] by V. Kobelsky, O. Viro, M. Ochaiai and O. Morikawa). Thus, the situation remained uncertain for the nets of genus two. (In the case of the nets of genus one, the answer is trivial; see Sec. 1, Ch. 3). In consequence, the author carried out a computing experiment especially for the nets of genus two; in the computer program, the revealed geometric mechanisms leading sometimes to wave destruction on the nets of the sphere of genus three and higher were taken into account.

It turned out that, despite the limited possibilities of a computer, the extended experiment did not discover a single disproof of the hypothesis on the existence of a wave on the nets of a sphere of genus two. Finally, in 1980, the hypothesis of Volodin and Fomenko was theoretically proved for all nets of a sphere of genus two in the work of T. Homma, M. Ochaiai and M. Takahashi [46].

THEOREM 2 (see [46]). *Each net of genus two of the standard sphere, different from the simplest, contains at least one wave. In particular, at least one Whitehead graph contains at least one separating vertex.*

The proof of this fact is non-trivial; therefore, we omit it. Note a curious result incidentally obtained in [46] and describing the structure of an arbitrary Whitehead graph of genus two, corresponding to a three-dimensional manifold. It turns out that each Whitehead graph is isomorphic to one of the plane graphs of Types I, II, and III, shown in Fig. 13, where by a, b, c, and d we

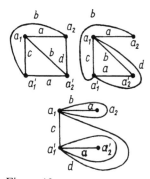

Figure 13

denote the numbers of "parallel edges" connecting corresponding pairs of the graph vertices.

Thus, the efficiency of an extremely simple and visual algorithm of recognizing the standard sphere in the class of Heegard diagrams of genus two, formulated by I. A. Volodin and Fomenko and tested primarily on a computer [40]. Viz., take an arbitrary net s and ascertain whether it contains at least one wave, i.e., whether there is a separating vertex in at least one of its Whitehead graphs. If there are no separating vertices on both graphs, then the original net determines a three-dimensional manifold not diffeomorphic to the standard, and we get the answer at the very first step. If there is at least one separating vertex, then we apply the operation of simplifying a net (of Whitehead graphs) of simple algorithmic character. As a result, we obtain a new net s' corresponding to the same manifold, and find out again whether there is a separating vertex. If it is absent, then the original manifold is not a sphere. If there is a separating vertex, then we simplify the net, etc. Thus, as a result of a finite number of steps, which does not exceed the number of edges of the original net, we stop either at the net which does not have a single wave (in this case, the original manifold is not a sphere) or at the simplest net s_0 corresponding to the canonical specification of the sphere. In the latter case, the original manifold is diffeomorphic to the sphere. This algorithm completes the known Haken algorithm.

3. On Solving the Four-Dimensional Poincaré Problem: Any Four-Dimensional Homotopy Sphere is Homeomorphic to the Standard Sphere

In 1904, the French mathematician H. Poincaré in his work *Analysis sitûs* suggested the following hypothesis: If a three-dimensional manifold M has the same integral one-dimensional homology as the standard three-dimensional sphere S^3, i.e., $H_1(M, Z) = 0$, then M is homeomorphic to the sphere S^3. This assumption has turned out to be incorrect. Poincaré constructed a three-dimensional manifold M_0 with a zero one-dimensional homology group, not homeomorphic to the sphere. The thing is that this manifold is not simply connected. Its fundamental group is of the form $\pi_1(M_0) = \langle a, b | (ab)^2 = a^3 = b^5 \rangle$. Three-dimensional manifolds with $H_1(M, Z) = 0$ are called three-dimensional homology spheres. The manifold constructed by Poincaré is sometimes called *the dodecahedral space*. See, e.g., [3, 34, 55] for the details.

Owing to the above circumstance, the hypothesis assumed its final form: If $\pi_1(M^3) = 0$, then the manifold M^3 is homeomorphic to the sphere S^3. By *Poincaré duality*, it follows easily that if $\pi_1(M^3) = 0$ then M^3 is homotopy-equivalent to S^3; therefore, the Poincaré hypothesis can be formulated as follows: if $M^3 \sim S^3$ ($\sim$ denotes homotopy equivalence), then $M^3 = S^3$. In this form, the hypothesis makes sense for manifolds of any dimensions; viz., if $M^n \sim S^n$, then $M^n = S^n$. Note that, for manifolds of dimension 4 and higher, the simple connectedness of a manifold does not mean at all that it is homotopy equivalent to the sphere.

In fact, accuracy is essential in formulating the Poincaré hypothesis. The class of the manifold as well as the sense of the equality $M^n = S^n$ should be specified. At present, three categories are most frequently used in the topology of manifolds: Top, of topological manifolds and homeomorphisms, PL, of piecewise linear manifolds (i.e., the manifolds regularly decomposed into simplices so that this triangulation is locally arranged as that of Euclidean space; see Ch. 1) and piecewise linear homeomorphisms (i.e., the homeomorphisms preserving the trangulation; see Ch. 1), and, finally, Diff, of smooth manifolds and diffeomorphisms. Thus, in each dimension n, it makes sense to consider three possible variants of Poincaré hypothesis $P_n(\text{CAT})$, where CAT denotes categories Top, PL or Diff.

Thus, the hypothesis $P_n(\text{CAT})$ is formulated in the following way: If M^n is a CAT manifold and $M^n \sim S^n$, then $M^n \overset{\text{CAT}}{=} S^n$, i.e., M^n is equivalent (homeomorphic, piecewise linearly homeomorphic, diffeomorphic) to the sphere S^n in the sense of CAT.

Here is a survey of the pre-1981 results connected with the Poincaré hypothesis. The survey is brief, and does not pretend to be exhaustive.

(a) There are relations between the hypotheses $P_n(\text{CAT})$.

1. If $n \neq 4$, then $P_n(\text{Top}) = P_n(\text{PL})$. See the works by Kirby and Siebenmann (1970) for $n \geqslant 5$, Mois (1952) for $n = 3$, and Page (1925) for $n = 2$. For $n = 1$, the statement is trivial.

2. If $n < 7$, then $P_n(\text{Diff}) = P_n(\text{PL})$.

(b) Point out dimensions for which the Poincaré hypothesis is valid.

1. The hypothesis $P_n(\text{PL})$ holds for $n \neq 3, 4$. See the works by Smale (1961) for $n \geqslant 5$. For $n = 2$, validity of the hypothesis follows from classification theory for two-dimensional surfaces.

2. The hypothesis $P_n(\text{Diff})$, generally speaking, does not hold for $n \geqslant 7$. Milnor (1956) proved the existence of non-standard smoothness on the spheres for $n \geqslant 7$.

This information can be arbitrarily schematized as shown in Fig. 14.

At the end of 1981, an American mathematician M. H. Freedman proved that the hypothesis $P_4(\text{Top})$ is valid, i.e., he placed a plus in the right circle labeled four in Fig. 14. He proved that a closed topological four-dimensional manifold M^4, homotopy-equivalent to the sphere S^4, is homeomorphic to it. In fact, he proved the *h-cobordism theorem* (see [3]) for four-dimensional manifolds. The Poincaré hypothesis easily follows from this theorem [56, 62].

Recall that a compact $(n+1)$-dimensional Pl-manifold W^{n+1} having the boundary composed of two components M^n_1 and M^n_2 is referred to as an

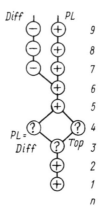

Figure 14

h-cobordism, if the film W, in deforming, is retracted onto both boundaries M_i^n, $i=1,2$, and if $\pi_1(W)=0$. The simplest example of h-cobordism is the direct product $M_1 \times I$. The Smale theorem states that if $n \geqslant 5$, then $W \overset{PL}{=} M_1 \times I$, where W is an h-cobordism. It follows, in particular, that the boundaries M_1 and M_2 of the h-cobordism are PL-homeomorphic.

The Poincaré PL-hypothesis for $n \geqslant 5$ from the Smale theorem can be derived in the following way. Let $M^n \sim S^n$. Remove a small open ball from the manifold. As a result, we obtain an n-dimensional manifold V^n. Consider $V^n \times I$, and remove the ball from this direct product. We have obtained an h-cobordism between two manifolds S^n and $\partial(V^n \times I)$. By the Smale theorem, this h-cobordism is trivial, and, therefore, $S^n \overset{PL}{=} \partial(V^n \times I)$. It remains to note that V^n lies in $\partial(V^n \times I)=S^n$ and is bounded there by the sphere, S^{n-1}. By the Schoenflies theorem, $V^n \overset{TOP}{=}$ standard n-dimensional ball. Hence, $M^n \overset{TOP}{=} S^n$, Q.E.D. (Fig. 15). The above argument can be specified to obtain $M^n \overset{PL}{=} S^n$.

Freedman proved that the theorem of h-cobordism holds for $n=4$, too, though in a weaker sense: If W^5 is a PL-h-cobordism, then $W^5 \overset{TOP}{=} M_1 \times I$.

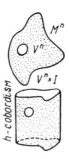

Figure 15

Here, the manifold W^5 is supposed to be compact. In fact, this theorem holds for a non-compact h-cobordism W^5 between non-compact manifolds. To this end, there should be an additional requirement for continuous deformations of the film W^5 to its boundaries M_i^4 to be proper (i.e., the inverse image of a compactum should be a compactum), and the film W^5 to be "simply connected at infinity". The latter means that the remote loops are contracted into a point, remaining "remote", i.e., the contracting homotopy should not affect the prescribed compactum, arbitrarily great it is. The hypothesis $P_4(\text{Top})$ easily follows from the non-compact h-cobordism theorem. If $M^4 \sim S^4$ is a PL-manifold, then the above argument is repeated in full. If M^4 is not a PL-manifold, then $M^4\backslash_*$ is contractible (here $*$ denotes a point); therefore, it is parallelizable. It follows from immersion theory that $M^4\backslash_*$ is immersed into the Euclidean space $\mathbf{R}^4$, and has, therefore, a PL-(or Diff-)structure. Repeating the above argument word for word, we obtain an h-cobordism of W^5 between S^4 and $\partial(V \times I)$, which is a PL-manifold outside a segment or, equivalently, outside a point. Now, we can remove the segment passing through this point and connecting the boundaries of the cobordism; after that, we have to apply the non-compact h-cobordism theorem to the remaining manifold.

We give a sketch of the proof of the h-cobordism theorem for four-dimensional manifolds, relying on the information contained in [3], [62]:

  (i) The Smale proof of the h-cobordism theorem in dimensions $n \geqslant 5$ is obtained in dimension 4 until the so-called Whitney trick is necessary;

  (ii) A detailed analysis of the Whitney trick reveals four reasons for this trick's inefficiency in dimension 4;

  (iii) The so-called Casson reduction enables us to cope with three of these difficulties by means of usual finitary topological methods, and, partially, with the fourth difficulty, applying a certain infinite process;

  (iv) The Freedman theorem of a *Casson handle* being homeomorphic to a real handle helps overcome the last difficulties. We describe briefly the subsequent steps in realizing this program.

The first step. The Smale proof of the h-cobordism theorem is as follows. First, an h-cobordism W^{n+1} is separated into handles, a small neighborhood of boundaries being distinguished a priori. By means of special operations with handles, we attempt to simplify this original partition, the final aim being to remove extra handles. We succeed in annihilating handles of indices $0, 1, \ldots, k-1$ (here, we assume that $n = 2k$), and, in virtue of duality, those of indices $n+1, n, \ldots,$ and $k+2$.

There remain handles of indices k and $k+1$ only. Applying this argument in the case $n = 4$, we obtain that only handles of indices 2 and 3 remain. Moreover, we can separate the handles of indices k and $k+1$ into algebraic complementary pairs. It means the following. A manifold M^4 which separates the handles of indices k and $k+1$ contains two collections of embedded spheres. The first collection $S = \{S_1^k, S_2^k, \ldots, S_m^k\}$ consists of the right-hand spheres (boundaries of separatrix disks of dimension $k+1$) embedded in the level surface $M^4 = \{f = 1/2\}$ for the corresponding smooth Morse function. These right-hand spheres correspond to handles of index k. The second collection $\tilde{S} = \{\tilde{S}_1^k, \tilde{S}_2^k, \ldots, \tilde{S}_m^k\}$ consists of left-hand spheres embedded in the same manifold $M^4 = \{f = 1/2\}$. These spheres are the boundaries of separatrix disks of dimension $k+1$ emanating from critical points of indices $k+1$. These points correspond to the handles of index $k+1$. We can assume that left-hand and right-hand spheres intersect transversally; in particular, at a finite number of points. The fact that these sphere algebraically complete each other means that the intersection number u of these spheres has the form $\mu(S_i^k, \tilde{S}_j^k) = \delta_{ij}$ (Fig. 16). If these two collections of spheres could complete each other geometrically and not algebraically, i.e., if the sphere S^k did not intersect the spheres $\tilde{S}_j^k$ with the numbers $j \neq i$, and intersected the sphere $\tilde{S}_i^k$ at one point exactly, then these pairs of handles which complete each other could be removed. In other words, a pair of such handles would cancel each other, and in their neighborhood, the film would acquire the form of the direct product. How the handles of complementary indices cancel each other if separatrix spheres intersect is described in [3] (Fig. 17). The Whitney trick is used to remove extra intersection points of separatrix spheres, i.e., to replace the spheres algebraically completing each other with those completing each other geometrically.

The second step. The *Whitney trick* is as follows. Given two intersection points of the spheres S and $\tilde{S}$, and let intersection numbers be $+1$ and -1 (Fig. 17(b)). The embedded two-dimensional Whitney disk spans the Whitney loop formed by two paths emanating from one intersection point to another on different spheres. Alone this disk, one sphere is deformed so that it contracts and slips off the other sphere by means of an isotopy (Fig. 17(b)). To perform this operations, the *Whitney loop* should be null-homotopic in the space $X = M \setminus (S \cup \tilde{S})$, i.e., in the complement to the spheres. This is deliberately provided by the following condition: $\pi_1(X) = 0$. If the ambient manifold dimension is greater than four, this condition is satisfied, since $\pi_1(M) = 0$ and

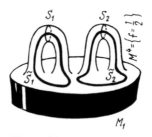

Figure 16

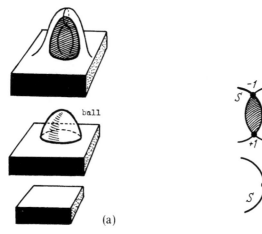

Figure 17

the codimension $S \cup \tilde{S}$ is great. In dimension 4, these conditions are not satisfied. This is the first reason by which the above Whitney trick is hampered.

If $\pi_1(X) = 0$, then the two-dimensional Whitney disk with singularities (singular disk) can span the above loop. In the case of a great codimensionality the disk can be embedded by being stirred slightly, theorems of general position being applied. In dimension 4, these arguments are not, generally speaking, enough. This is the second reason why the Whitney trick does not work. However, it is more convenient to call this reason the fourth for considerations which are listed below. To prove the theorem, we should make Whitney disks spanning different Whitney loops disjoint. In this case, the removal of extra intersection points can be performed simultaneously. Conversely, i.e., if Whitney disks intersect, then simple connectedness of the complement of X may be violated when point pairs are removed for the first time. Again, we can make the intersection of Whitney disk empty in higher dimensions though, generally speaking, prior methods do not enable us to do so in dimension 4. This is the third reason why the Whitney trick cannot be applied. Finally, the second reason lies in the fact that the Whitney disk should be well adjacent to both spheres. It is easily performed in higher dimensions though, in dimension 4, a certain integral obstruction arises which hampers good (zero) adjacency. This effect consists in the following. Let the Whitney disk be already embedded in the complement to spheres. Then its normal bundle (with two-dimensional fibers), restricted to a closed Whitney loop, has two trivializations. One trivialization is induced by a normal bundle having a disk as base space; therefore, trivially, it is a direct product. The second trivialization is determined by vector fields normal to the disk and touching the sphere S on one half of the Whitney loop and the sphere $\tilde{S}$ on its other half (Fig. 18). These two trivializations are different by a mapping of S^1 into $SO(2)$,

Figure 18

i.e., by an element of the group $\pi_1(SO(2)) = Z$. Thus, the reasons why the Whitney trick (in its prior meaning) does not work in dimension 4 are as follows:

(i) The non-simple connectedness of the complement to the spheres; (ii) Poor adjacency of the two-dimensional Whitney disk to the spheres; (iii) The intersection of different Whitney disks spanning different Whitney loops; (iv) The self-intersection of Whitney disks.

It turns out that these reasons can be removed. To overcome the first obstruction, the Casson trick is applied. It consists in the following. It can be seen from the problem that we can replace one family of spheres with isotopic families. It turns out that to annihilate the fundamental group of the complement (if it was non-trivial), isotopies of quite a special form suffice. A part of the sphere S is pulled into a "tongue", or "finger" along a path, and pierces the second sphere $\tilde{S}$, two new intersection points arising. The indices of these points are $+1$ and -1 (Fig. 19). It is remarkable that though our aim is to remove extraneous intersection points, the first step in solving this problem is adding new points, i.e., making the picture more complicated.

It is stated that the complement fundamental group is reduced when a pair of points is added. More exactly, there exists an equality $\tilde{\pi}_1 = \pi_1/w$, where $\tilde{\pi}_1$ is a new fundamental group, π_1 is a prior group, and w is a certain element of π_1.

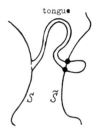

Figure 19

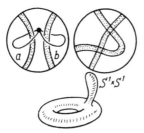

Figure 20

It is clear that it suffices to analyze developments in the small neighborhood of the point of the sphere $\tilde{S}$ near which the "tongue" pierces the spheres $\tilde{S}$ (Fig. 20). It is apparent that prior to the sphere being pierced by the "tongue", the complement to the two disjoint disks in the ball was homotopy-equivalent to the wedge of two circles, i.e., a figure of eight. The sphere being pierced by the "tongue", the complement to the two disjoint disks became homotopy-equivalent to gluing two tori along the coordinate axes. Thus, in piercing the sphere, the fundamental group of the complement with respect to the commutator w of paths a and b is factorized (Fig. 20). It is obvious that we completely annihilate the fundamental group of the complement after several such operations, since the π_1-group was perfect from the very beginning. It follows from $H_1(X) = 0$, which in its turn follows from meridians of the spheres being homologous to zero.

The removal of the second obstruction (see above) can be performed by more or less usual topological means. The details are omitted here.

To remove the third obstruction for Whitney trick application, we should first span one Whitney disk; then, by the Casson trick, let the "tongues" emanate from the disk, and make the complement to the disk simply connected. We let the subsequent disk span the subsequent Whitney loop in addition to the first disk, which is simply connected. This obviously enables us to make intersections of different Whitney disks empty.

The first three reasons for obstructing the Whitney trick being removed, we obtain some disjoint immersed Whitney disks with a good adjacency. For simplicity, consider the case where there is one disk. Distinguish a certain closed neighborhood of the disk, fibered into normal two-dimensional disks of small radius. Sometimes it is called a twisted handle. The twisted handle is completely determined by the number of positive and negative double self-intersection points of the disk. Topologically, this handle is homeomorphic to the connected sum of several copies of $S^1 \times D^2$. Each direct product arises as a result of the emergence of closed loops originating and ending at double points (Fig. 21). In the figure, such a handle is homeomorphic to $S^1 \times I$. Loops emanating from double points along one branch and returning along the other are referred to as *Whitney loops*. It is obvious that they can be driven onto the twisted handle boundary. Notice an

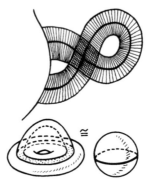

Figure 21

important observation: If Whitney loop is glued with the handle of index 2, we obtain a ball, i.e., a real handle. Similar operation in the three-dimensional case is shown in Fig. 21. To realize this sewing inside the manifold, two-dimensional disks should span Whitney loops embedded into the manifold and well-adjacent to the twisted handle, the disks being pairwise disjoint. As in the prior cases, all these conditions, except the condition for disks to be embeddable, can be satisfied. Glue Whitney loops with immersed two-dimensional disks and take their closed neighborhoods. Hence, we obtain the second layer of twisted handles. If Whitney circles could be glued with embedded disks inside the ambient manifold, satisfying the above conditions, then these twisted handles would transform into real. In this case, the original handle would become real too, though, in the generic case, we cannot achieve it yet. Therefore, we are to repeat this process and construct the third layer of twisted handles, etc. The union of such twisted handles (resulting from the described process) is referred to as *the Casson handle* (Fig. 22). Neglecting the lateral surface of the Casson handle, we obtain an open Casson handle. Freedman proved that any open Casson handle is homeomorphic to the direct product $D^2 \times \mathbf{R}^2$, i.e., a real handle.

The theorem of h-cobordism in dimension 4 follows immediately from this theorem. The point is that the axis $D^2 \times \{0\}$ of the handle $D^2 \times \mathbf{R}^2$ is the required; though, only topological Whitney disk embedded into the ambient manifold. It means that the Whitney trick works, but violates the smoothness in dimension 4. Further we obtain a usual proof of the theorem of h-cobordism and s-cobordism, too.

We state briefly the proof of the theorem on an open Casson handle being homeomorphic to the usual handle. An open subset G of the open Casson handle $\mathring{K}$ is constructed, which has the form shown in Fig. 23. Each pair of symmetrical segments of the type AB and A′B′ depicts a complement in S^3 to a "thick" Whitehead link in S^3, while branching is on the tori (Fig. 23(b)). In the standard handle $D^2 \times R^2$, an open subset G_1 is constructed (Fig. 24). Each

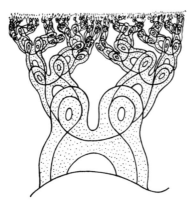

Figure 22

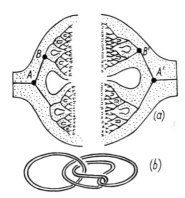

Figure 23

square is the product of a line segment and a solid torus lying in $D^2 \times \partial D^2 = D^2 \times S^1$, as shown in Fig. 24(b). This direct product is "pressed" into the manifold $D^2 \times R^2 = D^2 \times \text{Int } D^2$ as deep as required. It is easily seen that the nets G and G_1 are homeomorphic. This is a trial construction of the required homeomorphism of the Casson handle K with the real handle $\mathring{P} = D^2 \times R^2$.

The embedded disk (the dotted line in Fig. 24) spans the parallel of the torus, and each torus with the disk is contracted to a point. We obtain a certain space X and the mapping $g : \mathring{P} \to X$. The mapping $f : \mathring{K} \to X$ is constructed from the superposition on G and by contracting the complement components to G to corresponding points. Care should be taken of the mapping under construction to be continuous. We obtain the diagram $\mathring{K} \xrightarrow{f} X \xleftarrow{g} \mathring{P}$. The method developed by Bing enables us to prove that the mapping g can be replaced with a homeomorphism. The method developed by Freedman enables us to prove that the mapping f can be replaced by a homeomorphism, too. Let us dwell on the properties of g. Take a line segment in the square, and contract it to a point

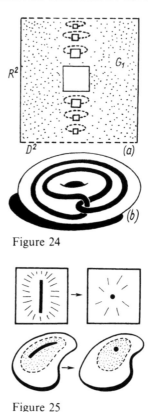

Figure 24

Figure 25

(Fig. 25). It is obvious that we obtain a square. Now, if in the handle P (and not in the square) a disk with a thicker boundary (i.e., a torus with a disk spanning a parallel) or several thicker disks (and not a line segment) are contracted to a point, then the handle P is inevitably obtained. The principle of Bing construction enables us to prove this statement for such thicker disks. In our case, this very situation is realized, the mapping g being replaced with a homeomorphism therethrough.

Let us dwell on the properties of the mapping f. Take a mapping of the sphere S^4 onto itself, which is an "almost-homeomorphism". To be more exact, under this mapping, the inverse image of each point (except one) is a point. The inverse image of one "singular" point is an arbitrary compactum Y. If a singular point A is enclosed with the sphere S^3, then its inverse image is a three-dimensional sphere bounding, by the Schoenf lies theorem, a four-dimensional cell. Therefore, the compactum Y can be compressed into an arbitrarily small cell. Such compacta are called *cellular* (Fig. 25). It is clear that a cellular compactum can be contracted to a point inside a sphere, thus replacing the original mapping with a homeomorphism. This argument holds for the Casson handle and the finite number of singular points. The method developed by Freedman enables us to extend the argument to the case of a countable nowhere dense set of singular points. This technique is due to the work by M. A. Shtanko [57]. This very situation is realized in our case; hence, the mapping f can be replaced by a homeomorphism. This completes the proof of the theorem.

The Poincaré hypothesis is closely connected with manifold classification; therefore, it is no wonder that the theorem of an h-cobordism between 4-dimensional manifolds enabled us to prove not only the Poincaré hypothesis in dimension 4 but also to classify simply connected four-dimensional manifolds. We dip briefly into this result, and specify the class of manifolds to be classified. These are closed simply connected topological almost smoothable four-dimensional manifolds. Being almost smoothable means that a manifold becomes smoothable after removing one point. As an equivalence relation, a homeomorphism is considered. Two algebraic invariants are necessary for classification.

The first invariant. This is a symmetric bilinear integral unimodular form defined by the intersection number on two-dimensional homology of the manifold, i.e., on the group $H_2(M^4)$. In other words, associating each pair of two-dimensional cycles with their intersection number, we obtain the required form. Therefore, a certain form L_m naturally corresponds to each four-dimensional manifold; e.g., the form U with a matrix $\begin{pmatrix} 0 & 1 \\ 1 & 0 \end{pmatrix}$ corresponds to the manifold $S^2 \times S^2$, and the form with a matrix (1), to the complex projective space CP^2.

The second invariant. This invariant denoted by κ_M is an obstruction to the introduction of a vector structure on the stable topological tangent bundle of the manifold M. This obstruction is in the group $H^4(M, Z_2) = Z_2$. In other words, $\kappa_M = 0$ if the stable tangent bundle of the manifold M admits vector reductions; otherwise, $\kappa_M = 1$.

To formulate classification theory conveniently, all four-dimensional manifolds are broken into two classes: even manifolds (with "even" form) and old manifolds (with "odd" form). The "evenness" of a form means that, on the diagonal of one (and then each of its matrices), there are even numbers; otherwise, there is at least one odd number.

The classification theorem of four-dimensional manifolds.

(1) *The correspondence $M \to L_M$ induces a bijection between the set of even manifolds (with the above equivalence relation) and even forms.* (2) *The correspondence $M \to (L_M; \kappa_M)$ induces a bijection between the set of odd manifolds and the set of pairs: (odd form, zero or one).*

The absence of the invariant κ in the even case is accounted for by its being expressed by L_M, i.e., it is not an independent parameter. To prove classification theorem, we should see that: (i) Manifolds with the same invariants are homeomorphic; (ii) Any values of invariants are realized.

Note that there is a generalization of the Novikov-Wall theorem on compact smooth four-dimensional manifolds with the same forms being h-cobordant to the non-compact case. Viz., the following statement is valid: If $L_{M_1} = L_{M_2}$ and, in the case of odd forms, $\kappa_{M_1} = \kappa_{M_2}$ then $M_1 \backslash_*$ is smoothly h-cobordant to $M_2 \backslash_*$ (in the non-compact sense, with "a good behavior at infinity".

By the h-cobordism theorem, manifolds M_1 and M_2 are homeomorphic in this case, which proves the first proposition of classification theorem.

To prove the second part of classification theorem, the realizability of an arbitrary form of the above type should be proved. To this end, we need (i) "the excision method" and (ii) the existence of "Rohlin Cap".

1. Let us describe the excision method. Given a four-dimensional manifold M^4, and let $M_1 = M \# (S^2 \times S^2)$. The equality $L_{M_1} = L_M \oplus U$ occurs. However, the converse also holds but in the topological sense; viz., if $L_{M_1} = L_M \oplus U$, then there exists such Top-manifold M^4 that $M_1 \overset{\text{Top}}{=} M \# (S^2 \times S^2)$. This proposition is proved in the following way. Realize two cycles a, b, giving the form U, by two immersed spheres with one generic point. For this purpose, first realize one cycle; then, by means of the above Casson trick, make the complement 1-connected, and after that realize the second cycle in this complement, which intersects the first at one point (Fig. 26). Further, just as in proving the h-cobordism theorem, we span Whitney loops by the above procedure, i.e., those with twisted handles. As a result, we obtain a pair of topologically standard handles glued to the ball, i.e., $(S^2 \times S^2) \backslash D^4$, where D^4 is a small disk. Removing the manifold $Q = ((S^2 \times S^2) \backslash$ small open disk) from the manifold M_1 and gluing the manifold boundary to the ball, we obtain the required manifold M (Fig. 26). The constructed manifold meets our requirements.

2. Describe the *Rohlin cap*. A *Rohlin cap* is a contractible almost smoothable manifold M_0^4, the boundary of which is the dodecahedral space M_0^3. Its

Figure 26

existence is proved in the following way. Take the direct product $M_0^3 \times I$. Recall that $\pi_1(M_0^3) = \langle x, y/(xy)^2 = x^3 = y^5 \rangle$. Perform a spherical surgery relative to the element x, removing its neighborhood homeomorphic to $S^1 \times D^3$ and gluing into the direct product $D^2 \times S^2$ instead, which results in a four-dimensional manifold V. It is easily seen that $H_2(V, Z) = Z \oplus Z$ holds with form U. One group Z is determined by the two-dimensional sphere linked to the loop x; the other, by the glued disk $D^2 \times (_*)$ and by the film reducing the homology of the loop x to zero inside the manifold M_0^3. Cutting out the carrier of U, we obtain a manifold W with boundary $M_0^3 \cup M_0^3$, having the homotopy type of $S^3 \times I$. Glue these manifolds W together into the "infinite tube" W_∞ which is homotopy-equivalent to $S^3 \times R^1$ (Fig. 27). Suppose that this tube can be halved by the sphere S^3 with a collar (i.e., with neighborhood which is the direct product of the sphere and a segment); see a dotted line in Fig. 27. Then, the required cap is constructed in the following way. Take part of the tube W_∞, contained between the sphere (dotted line in Fig. 27) and a sufficiently distant copy of the manifold M_0^3, non-intersecting the sphere S^3, and glue S^3 to the ball (Fig. 27). The proof that the sphere S^3 halving the tube exists is performed in the following way. First, the line 1 is removed from the tube W_∞, and a smooth disk D piercing it at one point (Fig. 27). Then $W_\infty \backslash 1$ is properly homotopy-equivalent to R^4, since $W_\infty \overset{R}{\sim} S^3 \times S^1$. By the h-cobordism theorem for non-compact manifolds (see above), $W_\infty \backslash 1$ is smoothly h-cobordant to R^4, and, therefore, $W_\infty \backslash 1 \overset{\text{Top}}{=} R^4$. Represent the space R^4 as $S^4 \backslash *$. Then, by one point compactification, we obtain the sphere S^4 again; from the disk without a point $D \backslash (D \cap 1)$, a disk $\tilde{D}$ in the sphere S^4 is obtained, which is locally flat everywhere except, probably, one added point. Here, we can apply the Chernavsky-Kirby theorem by which a three-dimensional manifold in the

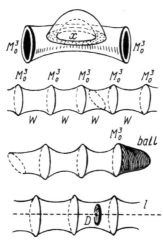

Figure 27

sphere S^4 cannot have isolated non-locally planar points. Therefore, the disk $\tilde{D}$ is locally flat, i.e., standard. Thus, it can be naturally complemented to a sphere. It means that the disk D in the tube W_∞ can be complemented to the required sphere, too. Q.E.D.

Now, we turn to the proof of the form realization theorem. First, realize the form E_8 with the matrix shown in Fig. 28. It is remarkable for its signature being equal to 8; therefore, it cannot be a priori realized by a smooth manifold; otherwise, it would contradict the Rohlin theorem by which the signature of an even smooth manifold must be divisible by 16.

Take eight copies of spaces of tangent bundles to the two-dimensional sphere S^2 onto two-dimensional disks, and paste them together by means of the so-called plumbing method according to the scheme in Fig. 29. On the boundary, a three-dimensional dodecahedral space M_0^3 is obtained [55]. The manifold form L is given by the matrix E_8, which follows from the manifold construction. It remains to reduce the manifold M_0^3 with the Rohlin cap.

The subsequent process of realization is obstructed by the absence of classification of all forms, though there is a classification of indefinite forms; viz., it is known that any even indefinite form is $kE_8 \oplus mU$, and any odd indefinite form is $k(+1) \oplus m(-1)$. These forms are realized easily. To this end, it suffices to take the connected sum of Rohlin manifolds with the manifolds $S^2 \times S^2$ in the first case, and that of manifolds $S^2 \times S^2$ in the first case, and that of manifolds $\pm CP^2$ in the second case. We now realize the definite forms.

2	1						
1	2	1					
	1	2	1				
		1	2	1			
			1	2	1		1
				1	2	1	
					1	2	
				1			2

Figure 28

Figure 29

Consider a certain form L. Adding the form U, we obtain the indefinite form $L \oplus U$. This form is realized as in the above case of indefinite forms, and then, by means of a convenient excision (see above), a realization of the form L is obtained.

In all considered cases where odd forms are realized, we obtained $\kappa = 0$, since the connected sum of projective spaces, i.e., $\#(\pm CP^2)$ is a smooth manifold, and when the carrier of the form U is cut out, the invariant κ does not change.

To realize the value of the invariant $\kappa = 1$ in the case of an odd form, the so-called "false projective space" $\pm CP^2$ should be constructed. This manifold is four-dimensional, and it has the form (± 1) and the value of the invariant $\kappa = 1$. The construction is performed in the following way. Take a four-dimensional ball, and draw a one-dimensional knot on its boundary, a trifolium. To this knot for the framing $+1$, a handle of index two is glued (Fig. 30). This operation results in the dodecahedral space [55], i.e., M^3, on the boundary of the obtained manifold. It should be glued to a Rohlin cap. The real projective space CP^2 is obtained by an analogous method, the standardly embedded circle being taken instead of the trifolium, and the sphere S^3 being obtained on the boundary. Now, it suffices to replace one of these manifolds (or their odd number) in the decomposition CP^2 with a false projective space. As a result, we obtain that $\kappa = 1$. Turn now to the problem of realizing forms by means of smooth manifolds.

Write out all four possibilities describing the types of forms. We give them as the following matrix:

	definite	indefinite
even	I	II
odd	III	IV

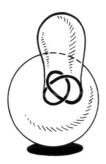

Figure 30

Relative to Types I and III, there is the Donaldson theorem: Of definite forms, only the form with unit matrix is realized by a smooth manifold. Relative to the form of Type II, the following is known. The form $E_8 \oplus E_8 \oplus U \oplus U \oplus U$ and, therefore, any form with number of forms U no less than $3/5$ of an even number of E_8 is realizable by a *Kummer surface*. Otherwise, the signature is not divisible by 16. All forms of Type IV (see the table above) are realized. Thus, we obtain the following matrix of realizability of forms of different types:

	definite	indefinite
even	—	part
odd	one series	all

The proof of the *Donaldson theorem* makes use of some up-to-date facts from partial differential equation theory; in particular, of the Young-Mills notion of self-dual connection, the instanton.

It turns out, then that the existence of non-standard smoothness on R^4 follows from the Donaldson theorem. The point is that if any smooth structure on R^4 were standard, then the algebraic decomposition $L_{M_1} = L \oplus U$ could be represented by the geometric decomposition $M_1 = M \# (S^2 \times S^2)$ not only in the topological but also in the smooth sense. This contradicts the Donaldson theorem, since three forms U could be pinched off the Kummer surface with form $E_8 \oplus E_8 \oplus U \oplus U \oplus U$ and a smooth manifold with form $E_8 \oplus E_8$ obtained. We describe this process in detail. Suppose that any smooth structure on R^4 is standard. Let us prove that the form E_8 can be pinched off in a smooth way. Recall that this procedure occurs as follows: In the manifold M_1, the cycles to be pinched off are realized by the immersed spheres intersecting at one point only. Then the killing of Whitney loops is done by pasting the twisted handles together. As a result, we obtain an open set H in the manifold M_1, homeomorphic to the punctured manifold $S^2 \times S^2$. This construction can be represented (in the smooth sense) in the standard manifold $S^2 \times S^2$ in the following way. Take the standard pair of spheres in the manifold $S^2 \times S^2$, introduce locally (by adding a self-intersecting two-dimensional sphere with one point of self-intersection in the four-dimensional space) as many positive and negative self-intersection paints as there are originally in the manifold M_1. Then, it is necessary to consider all two-dimensional disks reducing Whitney loops, embedded into the manifold $S^2 \times S^2$, and "spoil" these disks by adding small two-dimensional cross self-intersecting spheres (with one self-intersection point) to it so that the number of positive and negative self-intersection points will coincide with the number of corresponding points in the corresponding disks in the manifold M_1. As a result, we obtain an open set H_1 in the manifold $S^2 \times S^2$, diffeomorphic (!) to the set H. If the set H_1 contracted

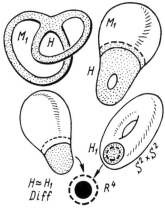

Figure 31

to a set homeomorphic to $(S^2 \times S^2)\backslash(\text{closed ball})$, is now removed from $S^2 \times S^2$, then an open subset homeomorphic to R^4 is obtained. Since the smoothness inherited from the manifold $S^2 \times S^2$ is standard on R^4, it can be extended to a smooth structure on S^4. It means that the smoothness on the set M_1 (contracted H) can be extended to the spanning disk, and a smooth manifold M such that $M_1 = M \# U$ in the smooth sense is obtained (Fig. 31). Thus, the proof is complete.

Similarly, the existence of a non-standard smooth structure on any smooth four-dimensional manifold M^4 without a point can be proved. To this end, we have to paste $S^2 \times S^2$ together in a smooth way, and cut it out roughly.

Summarizing, we have:

1. Any metrizable topological four-dimensional manifold, homotopy-equivalent to the standard sphere, is homeomorphic to this sphere. 2. Any topological four-dimensional manifold, properly homotopy-equivalent to R^4, is homeomorphic to R^4. 3. Any three-dimensional topological manifold whose integral homology is isomorphic to that of the standard three-dimensional sphere is the boundary of contractible topological four-dimensional manifold. 4. Closed simply connected topological almost smoothable manifolds (i.e., admitting a smooth structure on the complement to a point) admit a classification. 5. There exists a smooth four-dimensional manifold which is homeomorphic but not diffeomorphic to the standard four-dimensional Euclidean space.

In classification theorem, an important role is played by the form L_M of manifold M. If M is a closed simply connected manifold, then its integral homology and cohomology have the following form:

$$
\begin{array}{ccccccc}
i= & 0 & 1 & 2 & 3 & 4 \\
H_i(M)= & \mathbf{Z} & 0 & \mathbf{Z}^q & 0 & \mathbf{Z}. \\
H^i(M)= & \mathbf{Z} & 0 & \mathbf{Z}^q & 0 & \mathbf{Z}
\end{array}
$$

Here, q is the second Betti number of the manifold M. If a and b are elements of the group $H_2(M, Z)$, then their homological intersection number $a \cdot b = b \cdot a \in Z$ is defined. In virtue of Poincaré duality, $H_2(M) \approx H^2(M)$. Thus, there arises the symmetric bilinear non-degenerate form $L_M(a, b) = a \cdot b$; $L_M : H_2(M) \times H_2(M) \to Z$ which can be given by a symmetric integral unimodular matrix. This form completely determines the homological structure of a closed simply connected four-dimensional manifold.

CHAPTER IV/MINIMAL SURFACES

1. Simplest Properties of Minimal Surfaces

1. Plateau Physical Experiments and Methods for Obtaining Soap Films

Prior to the mathematical study of minimal surfaces, we give a brief description of their fundamental properties which can be demonstrated in the language of visual geometry. When the Belgian physicist, professor of physics and anatomy Joseph Plateau (1891—1883), started experimenting with soap films to examine their configurations, he could hardly believe that it would be the beginning of an important branch in science still developing rapidly and known today under the general name of the "Plateau problem". Some experiments carried out by Plateau are very simple and well-known to the reader: Probably, everyone amused oneself by blowing soap bubbles through a pipe or by constructing various soap films spanning a wire frame. It is well known that if a closed wire frame (e.g., homeomorphic to a circle) is dipped into soapy water and then taken out carefully, a beautiful iridescent soap film hangs from it. The film dimensions may be considerable; though the larger the film is, the sooner and easier it bursts due to gravity. If, on the contrary, the film dimensions are relatively small, then gravity may be neglected when some important properties of soap films are studied. This circumstance will be constantly used in the sequel. Minimal surfaces are a mathematical object modeling physical soap films fairly well. Conversely, many important properties of minimal surfaces are explicit in simple physical experiments with soap films. There are many branches of mathematics, which arose from concrete physical and applied problems. However, not all of them are as closely as the Plateau problem related to so many mathematical theories different in machinery. Within the framework of the theory of minimal surfaces, Lie groups, homology and cohomology theories, bordisms, etc., are interlinked.

In this section, we illustrate on simple examples some basic notions, techniques and results worked out in different periods of developing the Plateau problem. Since, in this domain, exact mathematical constructions require sometimes ramified, refined or non-trivial techniques, for the present, we confine ourselves to geometrically obvious constructions, omitting tedious computations. In particular, we to a considerable extent leave out generalizations to the multidimensional case, i.e., the so-called *multidimensional Plateau problem*. Some data are obtained from [11] and [19].

Considerable attention to the mathematical study of the properties of *soap films* was paid as far back as the 18th c. by Euler and Lagrange (problem of

finding the surface of the smallest area with given boundary). Exact solutions for some fairly special boundary frames were found by Riemann, Schwarz and Weierstrass already in the 19th c. The theory of minimal surfaces was given rise by considering two kinds of soap films: *soap bubbles* and soap films spanning a certain wire frame.

The first standard method of obtaining soap bubbles consists in sending air through a pipe with soapy water (Fig. 1). To stabilize bubbles which generally burst quickly, some glycerin is added to the water. Such soap films-bubbles are characterized by being formed and held by the internal pressure of the air inside the film. The spherical shape of a bubble is easily explained by the fact that this form provides for the smallest area of minimal surface for the fixed volume bounded by this film. The second method of obtaining soap bubbles is shown in Fig. 2. We should take a closed frame, e.g., a circle, dip it into soapy water, take it out, and quickly displace the frame in the direction orthogonal to the plane of the frame; for simplicity, the frame is considered to be a plane

Figure 1

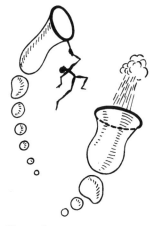

Figure 2

Figure 3

curve. Then the film sags under the pressure of a running air flow, and a few soap bubbles come off one by one. It is clear that the same effect can be achieved by directing the air flow to the soap film hanging from the fixed frame.

Another type of soap films is obtained when we take a wire frame out of the water, and fix it in space, trying to avoid any rough displacement. Then a stable soap film appears on the frame, generally without soap bubbles of described type in the film structure (they appear when air pressure inside and outside the film is different), the original wire frame being the boundary (Fig. 3). Note that, for the time being, the second type of soap films bounded by a frame is the main object studied by mathematical methods. This is due to the fact that, in many applied problems, a minimal surface is attached to a certain fixed boundary, as in membrane theory.

2. Physical Principles Underlying the Formation of Soap Films

The physical principle of forming soap bubbles and regulating their behavior as well as the local and global properties is extremely simple: The physical system retains its configuration provided it cannot easily alter it by taking a position with a smaller value of energy. In our case, the energy of a surface (soap film), often described in terms of surface tension, is caused by attraction forces between separate molecules and by these forces being unbalanced on the surface boundary. The presence of such unbalanced forces involves an interesting effect: A liquid film turns into an elastic surface tending to minimize its area and, therefore, surface energy per unit area. In this case, as before, we neglect gravitation (in the case of bubbles and films with boundary) and air pressure (in the case of films with boundary).

Let us see how the properties of a surface film change when some soap is added to the water. In Fig. 4, the surface of separation between the two media is represented. Black circles denote molecules of water. Some of them are in the air, since the liquid evaporates, and we neglect this effect. Arrows with two ends denote interaction attraction forces acting between polar molecules of water, which are characterized, as is known, by the charge distribution assymetry. It is obvious, that these interaction attraction forces give rise to the surface tension on the surface of separation between the media. In such a way,

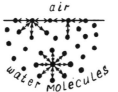

Figure 4

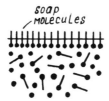

Figure 5

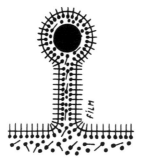

Figure 6

the properties of a surface film of a liquid are formed. In contrast to a molecule of water, molecules of soap are formed by long thin non-polar hydrocarbon chains with a polar oxygen group at one end of the chain. When added to the water, they tend to the surface and fill it in with an even layer (Fig. 5). Meanwhile, each molecule of soap is oriented on the surface with its non-polar end outside. Pushing the molecules of water inside the liquid, the molecules of soap thus lessen the surface tension on the surface of separation between the two media, apparently making the surface film more elastic, which manifests itself particularly at the moment of dipping and then extracting the thin wire frame from soapy water. In fact, when the wire, emerging from the liquid, reaches its surface, the latter bulges, spanning the frame (Fig. 6 shows its cross-section). This is due to the fact that the number of molecules of soap near the wire temporarily decreases per unit area of the film; therefore, the surface tension here increases. It is created by the molecules of water, breaking through from below and filling in the vacancies. This imparts to the surface film in the immediate vicinity of the wire extra elasticity and flexibility resulting in the appearance of a soap film when the frame is finally extracted from the water. The formation of a film hanging from the frame is shown in Fig. 6. As the frame is being raised, the surface film is not broken by the wire; it envelops the frame and trails behind, stretching from the liquid. As a result, we obtain a soap film spanning the frame. Forces of gravity restrict the dimensions of the minimal surface, and when some parts of the film are remote from the boundary frame, the film bursts, since surface tension forces are not sufficient for keeping the film in balance [11].

3. Extremely Properties of Soap Films and Minimality of Their Area. Properties of the Surface of Separation Between Two Media

It is owing to the described mechanism that the so-called minimal surface, i.e., the surface having the least possible area among all the surfaces with the same boundary, may serve as a mathematical model of a soap film. It is clear that the shape of the surface and its properties are to a considerable extent determined by the configuration of the boundary frame, i.e., the surface boundary. The notion of the least area was obviously introduced into geometry by Archimedes, who not only observed that the straight line realizes the shortest distance between two points, but also understood that the plane may be characterized in terms of minimal surface. These ideas arose from everyday practice; e.g., to make a drum, one had to stretch the parchment over a wooden frame. This allowed us to relate the idea of *surface tension* with the property of the surface to have minimal area. In general, the history of the development of true ideas of surface tension is very interesting. Here, we are going to touch upon only some stages of this process. Apparently, considerable progress was made by Robert Boyle who in 1676 took an interest in the shapes of drops of liquid [19]. He noticed that raindrops are roughly spherical, and decided to find the dependence of the shape of drops upon their size. To study how the drops behave for a sufficiently long time without decomposition, Boyle had to make a subtle experiment for those days. To this end, he poured two liquids into a glass vessel: a solution of K_2CO_3 (relatively heavy dense liquid, strong solution of potassium carbonate, or potash) and an alcohol (light liquid). When the liquids settled, the surface of separation between the two media was quite distinct (Fig. 7). Then, a drop of liquid of intermediate density and immiscibility with both liquids in the vessel was carefully placed on the surface of separation. Being immersed in alcohol and touching at its lower point the surface of K_2CO_3, the drop was in equilibrium. The third liquid (oil) was chosen, so that it did not wet the surface of separation, i.e., the drop neither spread over the surface of separation, nor had the shape shown in Fig. 7(b). Then, drops of different sizes were placed on the

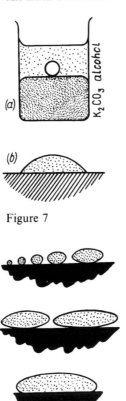

Figure 7

Figure 8

surface of separation; first, small ones; then, their size was gradually increased. It turned out that as the size increased, the shape of drops underwent a change (Fig. 8). Visually, flattening out becomes obvious when the drops are about the size of peas. Naturally, we now know that drops (even if they are small) flatten out in any case if there is gravity. However, to observe this phenomenon, high precision instruments were needed which were not at Boyle's disposal in 1676. In addition, it should be noted that contemporary ideas of drops shape association with minimum energy began developing only in the 17th c. Processes of that kind were mostly explained by phlogiston, which made the understanding of true mechanisms determining the shape of drops more difficult. In fact, the shape of a drop minimizes the combination of surface energy and force of gravity, which considerably complicates the mathematical aspect of the problem when drop size is small. In 1751, Segner understood this phenomenon and came to the conclusion that, to study surface tension accurately, he should try to eliminate the influence of gravity as far as possible. He partly succeeded in it, examining free-falling drops as well as those placed into a liquid with the same density. The immiscibility of the drop and the ambient liquid has been provided beforehand. In particular, he showed that the sphere has minimum area among all closed surfaces bounding fixed volume. Apparently, Segner was the first to understand the true role of surface tension in all these processes. However, the development of these ideas, including the idea of surface energy, we considerably hindered by the prevailing idea of phlogiston still unwilling to surrender [19].

4. The Surface of Separation between Two Media Which Are in Equilibrium Is the Surface of Constant Mean Curvature

An important step in understanding intrinsic geometry of the surfaces of separation between media was made by Poisson in 1828 when he showed that the surface of separation between two media, which is in equilibrium (provided that we neglect gravity), is the surface of constant mean curvature. We will dwell on this important result, expounding it in modern terminology and using concepts of modern differential geometry. It may be useful for the reader to be a priori familiar with some basic properties of the second fundamental form of two-dimensional surface embedded into a three-dimensional Euclidean space. See the corresponding facts in, e.g., [1, 5].

Let M^2 be a two-dimensional smooth surface in R^3, where R^3 is ascribed to Cartesian coordinates x, y, z. Let the surface be given by the radius vector $r = r(u, v) = (x(u, v), y(u, v), z(u, v))$, where parameters u and v are transformed in a certain domain on the Euclidean plane, determining regular coordinates in the neighborhood of a point P on the surface. Let $n(u, v)$ be the normal unit vector to the surface point $P = P(u, v)$. Consider a square symmetrical matrix Q of the numbers $q_{ij} = \langle r_{u_i, u_j}, n \rangle$, where r_{u_i, u_j} are second partial derivatives in the variables u and v (denoted here by u_1 and u_2, respectively); $\langle , \rangle$ denotes the Euclidean scalar product in R^3. If a and b are tangent vectors to the surface at the point P, then the expression $Q(a, b) = \sum_{i,j} q_{ij} a_i b_j$, where $a = (a_1, a_2)$, $b = (b_1, b_2)$, is a bilinear form determined on the tangential plane to the surface and referred to as the *second fundamental form* of the surface. It is clear that the second fundamental form depends on the method of surface embedding; the surface being deformed, the form will, generally speaking, change. In the two-dimensional case which is of interest to us now, the second fundamental form is generally written as $L du^2 + 2M du dv + N dv^2$, where $L = L(u, v) = \langle r_{uu}, n \rangle$, $M = M(u, v) = \langle r_{uv}, n \rangle$, $N = N(u, v) = \langle r_{vv}, n \rangle$. Here, L, M, N are smooth functions of the coordinates on the surface. In addition to the second fundamental form, consider the first fundamental form on the surface, determining the Riemannian metric induced on the surface by the underlying Euclidean metric. This form is usually denoted by $E du^2 + 2F du dv + G dv^2$, where $E = \langle r_u, r_u \rangle$, $F = \langle r_u, r_v \rangle$, $G = \langle r_v, r_v \rangle$. Denote the matrices of these forms by $Q = \begin{pmatrix} L & M \\ M & N \end{pmatrix}$ and $A = \begin{pmatrix} E & F \\ F & G \end{pmatrix}$, respectively. Then (see [1, 5]) two consecutive scalar functions $K = \det A^{-1}Q$ and $H = \operatorname{Spur} A^{-1}Q$ are important geometrical characteristics of the surface, called Gaussian and mean curvature, respectively. At present, mean curvature is of particular interest, since it is closely related to the geometry of the surface of separation between two media. By definition of mean curvature, $H = \dfrac{GL - 2MF + EN}{EG - F^2}$. It turns out that this expression can be essentially simplified by picking out local coordinates, which enables us to associate mean curvature with the surface tension on the surface of separation between the media. Recall that curvilinear regular coordinates u, v on the surface M are said to be *conformal* (or *isothermal*) if the induced Riemannian metric $ds^2 = Edu^2 + 2Fdudv + Gdv^2$ in them is diagonal, i.e., $E = G$, $F = 0$, and $ds^2 = E(du^2 + dv^2)$. It appears that if the metric on the surface is real-analytic, then there always exist local conformal coordinates in the vicinity of each point [1, 5]. In other words, for any point $P \in M$, there exists a neighborhood $U(P)$ such that we can introduce the coordinates p and q (which are real-analytic functions of the original coordinates) in which the metric ds^2 has the form $\lambda(p, q)(dp^2 + dq^2)$. We can assume the surfaces of separation between the media to be real-analytic at all interior singular points; therefore, without loss of generality, we can assume the existence of conformal coordinates in the vicinity of each regular point.

Thus, let u and v be conformal coordinates in a certain neighborhood of a point P on a surface M in R^3. The matrix of the first form may be considered to be unit at the point P. At one point, this can always be done by the corresponding orthogonal transformation of conformal coordinates. In particular, at the point P, we have $EG - F^2 = 1$. In virtue of the conformity of the coordinates, we obtain $E = G$, $F = 0$ in the vicinity of P; therefore, $H = (L + N)/E$. Mean curvature at the point P is even simpler: $H(P) = L + N$, since $E(P) = 1$. Recall the explicit form of the coefficients L and N. We obtain $H = E^{-1}(\langle r_{uu}, n\rangle + \langle r_{vv}, n\rangle) = E^{-1}\langle r_{uu} + r_{vv}, n\rangle = E^{-1}\langle \Delta r, n\rangle$, were $\Delta = \dfrac{\partial^2}{\partial u^2} + \dfrac{\partial^2}{\partial v^2}$ is the *Laplace operator*. Simplify the expression for mean curvature still further. Consider the tangent plane to the surface at the point P, pick out Cartesian coordinates x, y in it (Fig. 9), and direct the third coordinate z along the normal to the surface. Then, locally, the surface near the point P is written by means of radius vector $r(x, y) = (x, y, f(x, y))$, where $f(x, y)$ is a smooth function whose graph specifies the surface. Then we obtain $H(P) = \langle r_{xx} + r_{yy}, n\rangle$. Note that x, y must not necessarily be conformal on the surface, but they are orthogonal at the point P; therefore, $E(P) = G(P) = 1$, $F(P) = 0$. Since relative to the coordinates x, y, z the normal vector $n(P)$ is denoted by $n(P) = (0, 0, 1)$, then $H(P) = \Delta f = \dfrac{\partial^2 f}{\partial x^2} + \dfrac{\partial^2 f}{\partial y^2}$. For us, it is important

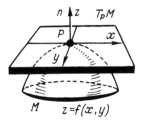

Figure 9

that the graph of the function f gives (locally) the surface M in the neighborhood of the point P. To associate mean curvature with surface tension, recall one more interpretation of the Laplace operator.

Let $f(x_1, \ldots, x_n)$ be a smooth function on R^n with respect to Cartesian coordinates $x_1, \ldots, x_n$. Define ρ-local averaging of the function f on R^n by $F_\rho f(x) = \dfrac{1}{\gamma_\rho} \cdot \int_{S_\rho} f$, where S_ρ is Euclidean sphere in R^n of radius ρ with center at the point $x = (x_1, \ldots, x_n)$, and γ_ρ is the volume of the sphere S_ρ. In other words, we average the original function f along the boundary of a spherical neighborhood of radius ρ with center at the point x. If the function is given on the plane, then S_ρ is a circle, and γ_ρ is its perimeter, $\gamma_\rho = 2\pi\rho$. Define now the "deviation of the function from its ρ-local averaging", putting $\Delta_\rho f(x) = f(x) - F_\rho f(x)$. Passing the limit as $\rho \to 0$, we can construct the function $g(x) = \lim_{\rho \to 0} \Delta_\rho f(x)$. It is easily verified (we leave verification to the reader) that the function $g(x)$ coincides up to a constant multiplier with the function $\Delta f(x)$. The function $\lim_{\rho \to 0} F_\rho f(x) = Ff(x)$ can be called the *local averaging* of the function f. Then $\Delta f(x) = f(x) - Ff(x)$, i.e., the value of the Laplace operator on the function f is equal to the deviation of the original function f from its local averaging Ff. Such interpretation of the Laplace operator is useful from many standpoints. For example, if f is a *harmonic function*, i.e., $\Delta f = 0$, then it coincides with its local averaging. In particular, a harmonic function does not attain a strict local maximum at any interior point of the domain where this function is defined (maximum is attained on the boundary).

Returning to the analysis of mean curvature, we see that the curvature $H(P) = f_{xx} + f_{yy} = \Delta f$ may be interpreted as the deviation of radius vector of the surface from its local averaging in the direction of the normal. Recall that radius vector is completely determined by the function $f(x, y)$. Suppose that the surface is the surface of separation between two media, on which there are attraction forces between nearby points as in the case of molecules of water in the above example. To make it clear, assume that we have to do with a soap film. Then we can write out a chain of equalities: The pressure acting on the surface at the point P is equal to the projection onto the normal to the surface of the resultant of all local attraction forces between the points near to the point P, which, in turn, is equal (up to a certain constant multiplier which does not depend on the point) to the deviation of radius vector from its local averaging (deviation in the direction of the normal). Thus, the pressure at the point P is equal to a certain constant multiplied by mean curvature of the surface at the given point. The constant resulting from this analysis, i.e., the coefficient of mean curvature, is called surface tension. Hence, the above result follows, viz., if a two-dimensional surface is either a soap bubble, a system of such bubbles, the surface of separation between two liquids of the same density (in which case gravity may be neglected) or a soap film spanning a wire frame, then the pressure on two sides of the surface is a constant function, i.e., it does not depend on the point provided the surface is in equilibrium. Therefore, mean curvature is constant. Q.E.D.

This accounts for a spherical shape of soap bubbles, which they acquire during a free fall. In this case, gravity can be neglected. Here, air pressure inside the sphere exceeds outside pressure, and equilibrium of the film results from surface tension forces which stabilize the spherical film. Particularly interesting is the case of a soap film spanning a wire frame. Here, there is no pressure difference on the two sides of the film. Therefore, the pressure on one side of the film coincides with the pressure on the other side in the vicinity of each point on the surface. Thus, the resultant force is zero (we neglect gravity); therefore, mean curvature is zero.

5. Soap Films of Constant Positive Curvature and Constant Zero Curvature

On finding out that soap films, i.e., "liquid surfaces", are surfaces of constant mean curvature, Poisson, naturally, raised the question of a full description of such surfaces. Since the problem of a possible shape of drops of a liquid had not yet been solved, at first, the researchers concentrated on the case of positive mean curvature. It is clear that the standard Euclidean sphere is such a surface; therefore, free falling soap bubbles do acquire a spherical shape. But it remained uncertain whether there are other closed (i.e., without boundary) surfaces of constant positive curvature. Naturally, numerous physical experiments with soap bubbles convince us quickly that the sphere is the only possible surface of constant positive curvature in the class of smooth closed surfaces, though a mathematical proof of this fact requires some effort. Poisson proved that, in the class of spheroids sufficiently close to the sphere and which are surfaces of revolution (i.e., slightly contracted or elongated along a certain axis), the sphere is the only surface with constant mean curvature. This information was practically unique up to 1853, when Jellet proved that, among closed two-dimensional star-shaped surfaces (i.e., each point of which is seen from a certain interior point, the "center" of the surface), the sphere is the only surface with constant mean positive curvature. Finally, in our century, it was proved that a compact closed two-dimensional surface of constant positive curvature, having no self-intersections, is the standard sphere. Note that we speak about surfaces embedded or immersed in R^3. In particular, the second form and mean curvature are defined only for the surfaces immersed in a certain ambient space.

Thus, we have singled out two main cases: If a soap film encloses a space with pressure inside greater than outside, then the surface is of constant positive mean curvature; otherwise, if it spans a wire frame, then there is no pressure difference, and mean curvature is zero. The first basic experiments made by Plateau were dedicated to the study of the surface of separation between liquids, the experiments being carried out with immiscible liquids of equal density. One liquid was immersed into another, which resulted in the formation of drops bounded by the surfaces of a constant positive curvature. Experiments with surfaces of zero curvature began later [19].

In the sequel, we shall concentrate on the case of soap films with boundary. These films admit, therefore, three equivalent descriptions: (a) surfaces of zero mean curvature; (b) surfaces of minimal area, minimal surfaces; (c) surfaces at each interior regular point of which principal curvatures are of the same module, but opposite of sign. The latter characteristic follows from $H = \lambda_1 + \lambda_2$, where λ_1 and λ_2 are principal curvatures.

6. Stable and Unstable Surfaces

Even the simplest experiments show that different surfaces of separation between two media respond to perturbations in different ways. Some of them resist destruction and turn out to be stable; others break up immediately, and they are unstable. Here, an important effect occurs, illustrated on the level of scalar functions on a smooth manifold. Recall that if $f(x)$ is a smooth function on a finite-dimensional manifold M, then critical points, i.e., in which grad $f = 0$, play a great role in the study of function behavior. These points may be of several kinds. First, critical are local minimum and maximum points of the function. In addition, there are the so-called saddle points. A typical form of such a point is schematically represented in Fig. 10. They are

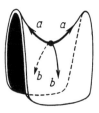

Figure 10

characterized by two kinds of directions: Along the directions *a* (Fig. 10) the function strictly increases, while along the orthogonal directions *b* it strictly decreases. The number of independent directions along which the function strictly decreases is generally called the *index of a critical point*. The points of non-degenerate local maxima have maximum index. The same sort of picture is observed in the case where the function (functional) is defined on the infinite-dimensional space. Such a space with a correct choice of its definition not to be sharpened here is the set of all two-dimensional surfaces, e.g., spanning wire frames in a three-dimensional space. Then the area functional associates each such surface with its area. If a two-dimensional film is in equilibrium, it means that, viewed as the point of infinite-dimensional space, it is critical for the area functional. But, just as in the finite-dimensional case, this equilibrium position may be of several kinds: the local minimum, the local maximum, and the saddle. In the last case, sufficiently small perturbations of the film exist, which decrease its area. When stable films are physically realized, we obtain that, in the case of a local minimum of the area functional, the film is stable, i.e., it resists small perturbations and tends to change to the initial equilibrium position; whereas in the case of the saddle, the film is unstable, and arbitrarily small amplitude perturbations arise, which decrease the area of the film. The film is quickly deformed, and it acquires a new configuration corresponding to a smaller value of energy, the new configuration being essentially different from the original from the topological standpoint (see the examples below). The "infinite-dimensionality" of the space of all surfaces also means that there exist infinitely many various independent directions of film perturbations.

7. Plateau Experiments in Stability of a Liquid Column

Plateau carried out a series of experiments with liquids. We have already dwelt on some of them. Here, we describe briefly another series of his experiments. Between two metal disks of the same radius, the centers of which lie on the same line orthogonal to the plane of the disks, Plateau would realize a liquid column in the form of a right circular cylinder made up entirely of liquid (Fig. 11). Since the cylinder boundary is the surface of constant positive mean curvature, theoretically such a liquid column is a critical point in the space of all the "surfaces". You should not confuse this case with the case of closed (!) compact smooth surfaces of constant mean positive curvature (as we know already, such a surface is a sphere). The cylinder is not a compact surface if it is infinitely extended in both directions. If it is bounded by two boundaries-circles (disks), it is not a closed (i.e., without boundary) smooth manifold. The properties of a cylindrical liquid column considerably depend on its height. In real physical experiments, a liquid column which is "not very high" is realized

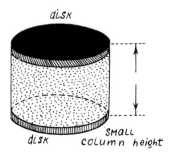

Figure 11

between two parallel disks. If two disks are considerably removed from each other, the liquid column breaks up. It is very interesting to see it breaking up when two disks, remaining parallel, are gradually moving away from each other, stretching the liquid column vertically. It turns out that the water column is stable only in the case where its height does not exceed three times the boundary disk diameter. If the column height approaches this value (approximately 3.1 times the diameter) and then exceeds it, the column begins breaking up. If the column height is increased gradually, the column behavior near the critical value of its height can be observed visually. In this case, the column deformation occurs sufficiently slow, which enables us to show the surgery of the column in Fig. 12. Thus, low cylinders are stable, and excessively high cylinders are unstable. The critical height is approximately three times the disk diameter. Indeed, these experiments are carried out with the disks of sufficiently small diameter so that surface tension forces can successfully compete with the force of gravity, acting on the liquid column.

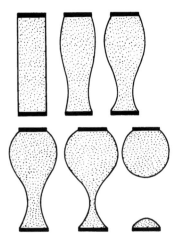

Figure 12

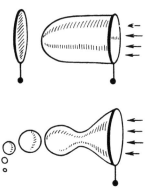

Figure 13

Decomposition process of unstable column, i.e., the process of its qualitative surgery, can be demonstrated by means of another real experiment performed as far back as the 18th c. Consider a wire frame in the form of a regular flat circle, and, placing it in soapy water, let a disk, i.e., soap film, span it. Then quickly move the frame in the direction orthogonal to its plane, i.e., orthogonal to the soap film (Fig. 13). Under the action of a running air flow, the soap film sags, and at first acquires (roughly) the shape of a right circular cylinder (this occurs at least in the immediate vicinity of the frame) though then this cylindrical shape decomposes quickly. Constrictions appear on it; the film undergoes a surgery, and soap bubbles of almost spherical shape start coming off the film. Obviously, the same process can be observed by fixing the frame in space and directing at it an air flow orthogonal to the plane of the frame. A cylindrical liquid column was the first critical surface the stability of which was analytically proved by Plateau. Then Beer suggested that all surfaces of constant mean curvature should be local minima of the area functional. Plateau proved that this assumption is not valid. He considered a cylinder of a height greater than critical. From the mathematical standpoint, such a cylinder is still a critical point (surface) for the areal functional in the space of all surfaces. First, Plateau proved analytically that such a cylinder has locally the least possible area in a special class of all perturbations of this cylinder, preserving the area of each horizontal section. In other words, this class of perturbations is characterized by the absence of the vertical displacement of the liquid. The horizontal displacement of the liquid, preserving the area of any plane section orthogonal to the primal axis of the cylinder, is admitted. In other words, if we perturb the cylinder so that it curves slightly while the plane sections preserve their area, the perturbed cylinder has a greater area than the original. Recall that if the cylinder height is smaller than critical, any (!) perturbation of the cylinder increases its area, i.e., in this case, the original surface is the point of local minimum for the areal functional in the space of all surfaces. If the height is greater than critical, then the cylinder is a

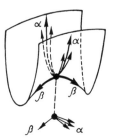

Figure 14

saddle point in the space of all surfaces. Meanwhile, there exist, as before, infinitely many independent directions α of perturbations of this critical surface in the space of all surfaces increasing its area. The infinity of the number of directions along which the critical surface is the minimum of the areal functional follows from infinitely many area-preserving deformations α.

On the other hand, if the height of the cylinder is greater than critical, then there exists at least one direction β in which the perturbation occurs, and the area begins decreasing. As we know already, to produce such perturbation, it is sufficient to set the liquid in vertical motion resulting in swelling one part of the cylinder, whereas the other part constricts (Fig. 12). The saddle is schematically shown in Fig. 14. Moreover, Plateau experimentally established the saddle nature of a critical surface in the case where the height of the cylinder exceeds three times the diameter of the boundary disk. Between two disks, he realized a liquid column of the height slightly greater than critical. The column began slowly deforming in vertical direction, while remaining a rotationally symmetric surface. At this moment, the liquid column was tapped with a thin glass stick in a horizontal position, i.e., the perturbation preserving the area of plane horizontal sections was realized physically. The column would deflect slightly, and regain its axially symmetric position almost instantaneously, meanwhile deforming slowly in the vertical direction. This fact points out the stability of the surface with respect to the perturbations of the above type α (Fig. 15). The deformation in the vertical direction is realized by the perturbation of type β. Curiously, decomposition of a sufficiently long liquid column is wave-like. To demonstrate this process visually, we can take a thin thread covered with a liquid cylindrical layer so that the thread is the axis of this cylinder. Then, the surgery of this cylinder occurs, shown in Fig. 16. The cylinder becomes wave-like, then each bulge becomes a sphere strung on the same thread. It is clear that this is the final result of the action of surface tension forces and instability of long liquid cylinders. Indeed, the process shown in Fig. 16 is a certain simplification of the real picture which is more complex, though, at the end, we obtain a necklace of spherical drops strung on the original thread just the same.

This particular process was studied in detail, since the problem of coating a

Figure 15

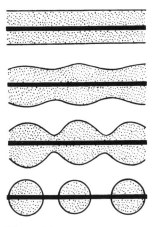

Figure 16

thin conductor (wire) with a cylindrical uniform layer of isolation arose (and was solved) in electrical engineering where it was important to learn to stabilize a long cylindrical layer enveloping the wire which goes axially in the cylinder (T. Poston, [19]).

8. Physical Realization of a *Helicoid*

In 1842, Catalan proved that the plane and helicoid are unique complete ruled surfaces of zero mean curvature. The helicoid is obtained as a result of the composition of a straight line: translational motion with constant speed and rotary motion with constant angular velocity in the plane orthogonal to the translation vector (Fig. 17). In other words, a line intersecting the second line at right angles should be taken and moved uniformly along the second line, rotating uniformly. This surface is non-compact. If we confine ourselves to a line segment instead of the whole line, we obtain the surface shown in Fig. 18.

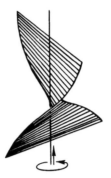

Figure 17

Figure 18

In coordinates (x, y, z) and (r, t), it is given by radius vector $x = r \cos \omega t$, $y = r \sin \omega t$; $z = at$, where a and ω are translational motion speed and rotary motion velocity, respectively. Plateau realized "half" the helicoid, using a wire spiral wound around a straight axis (Fig. 19). Nevertheless, an attempt to realize a complete helicoid by means of a film spanning a wire frame meets with some difficulties. Indeed, we can get part of the complete helicoid by putting together two copies of the film shown in Fig. 19. However, under this approach, we obtain that, in addition to the boundary frame composed of two helices with common axis, to realize a soap film, this axis should be added; e.g., in the form of a film going along the axis of the whole structure (Fig. 20). This thread stabilizes the film, and obstructs its restructuring into minimal surface which is not a helicoid any more. Indeed, if this axial thread is removed and only two boundary helices remain, then, in the case where the step of the

Figure 19

Figure 20

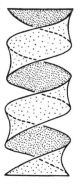

Figure 21

helices is not very large, i.e., when the helix does many turns at a fairly small distance along the axis, the helicoid becomes the surface shown in Fig. 21. It is apparent that this surface is not ruled, i.e., it does not consist of line segments. Attempts fail to realize part of a right helicoid as a soap film spanning a closed wire frame in the form of two coaxial helices with a small step. Meanwhile, it was supposed that the frame should not contain any additional wire segments which are strictly in the film and are not the film boundary. Here lies the important difference between the right helicoid and above surfaces. In fact, the helicoid is a non-compact surface existing (from the mathematical standpoint) independently of any boundary frame. In real experiments, we can realize only the soap films spanning some or other compact wire frames. Adding this condition to the realization of a helicoid, we encounter the above difficulties. If we abandon the attempt to realize the helicoid with small step, and aim at realizing the helicoid with a sufficiently large step of boundary helices, then it will not be difficult to construct such a film. To this end, it is necessary to consider a frame composed of two coaxial helices. It is clear that with a sufficiently large step this frame bounds a soap film, a helicoid (Fig. 22). The proof that the film structure is ruled can be given in the following visual way. It is obvious that the helicoid in Fig. 22 is obtained from the minimal surface in Fig. 21 by stretching boundary helices in the direction of the axis. As the step of the helix increases, the soap film is deformed. Fix in space the axis of the helices and the direction of our viewing the film. We observe the deformation of part of the film h which is in immediate vicinity of the intersection point of the helices (Fig. 23). Projected onto the eye retina, this part of the film has the form of a curvilinear triangle which, obviously, decreases as the helices stretch. At some moment, this triangle contracts to a point, and we obtain the film shown

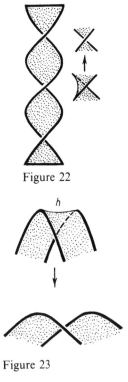

Figure 22

Figure 23

in Fig. 22. Since the projection of the film has the form given in Fig. 23, the helices run; e.g., from left to right when the film is rotated about its axis. The points of helices where their projections meet move in the same direction. Since the part of the film, projected (on the eye retina) into each of these points, is a rectilinear segment joining a point on the nearest helix to a point on the other helix, the whole film is composed of rectilinear segments orthogonal to the axis, which proves that the film is ruled and coincident with the helicoid.

9. Physical Realization of a Catenoid and Its Surgery with Changing Boundary Frame

The popular example of a minimal surface is obtained in the following way. Consider a minimal surface of rotation, generated by rotating the curve given by the equation $y = a \cdot \cosh \dfrac{x}{a}$ about the Ox-axis, where a is a non-zero constant. It is well known that this curve coincides in form with heavy sagging chain fixed at two points (Fig. 24). The force of gravity is along the Oy-axis. Direct computation shows that the mean curvature of the obtained surface of revolution is zero: $H = \lambda_1 + \lambda_2 = \dfrac{1}{y\sqrt{1+(y')^2}} - \dfrac{y''}{(1+(y')^2)^{3/2}} = 0$. Thus, we have obtained a minimal surface referred to as a *catenoid*. It can be regarded as a soap film spanning two coaxial circles in parallel planes. If we rewrite the catenary curve equation as $\dfrac{y}{a} = \cosh \dfrac{x}{a}$, then we obtain that all catenary curves are similar, and, therefore, all catenoids are equivalent in the sense that they can be carried into each other by the motion of a three-dimensional space and changing scale. In this sense, catenary curves are similar to parabolas also

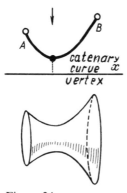

Figure 24

equivalent under motions on the plane and scale change (in contrast to hyperbolas and ellipses). A catenoid is obtained by rotating a catenary curve only in the case where this curve is at a certain distance from the axis of rotation. If the catenary curve is displaced, then, generally speaking, the surface of rotation to be obtained is no more minimal. Thus, o.g., if a catenary curve with small sag is nearly tangent to the Ox-axis, then the lower point of the curve, nearest to the curve, describes a circle of small radius when the catenary curve is rotated. Therefore, one of the curvature radii is very small, and the second is great; hence, the sum of principal curvatures at the points on the gorge circle of the surface of rotation is different from zero.

Consider how the catenoid structure depends on the size of the boundary frame. For simplicity, consider two coaxial circles of the same radius r, and let h be the distance between parallel planes containing them (Fig. 25). We assume the radius to be fixed. Introduce one more parameter a, the catenoid gorge radius, i.e., the distance to the axis of rotation from the points of the catenoid, nearest to it, which form the circle of radius a. It is clear that the parameters a and h are connected with each other. We can assume h to be a smooth function of a. Let us describe the qualitative characteristics of the function $h(a)$. The nature of the dependence of h on a is determined by the above circumstance: Since mean curvature of the catenoid is zero, the osculating circle radius at the vertex of catenary curve (at the point nearest to the axis of rotation) is equal to the radius of the circle described by the vertex of the curve rotated about the axis, i.e., a. Therefore, the smaller a, the greater the curvature of the catenary curve at its vertex. The parameter a transforms on the segment from zero to infinity. If $a > r$, then it is clear that no minimal surface spanning two circles exists (Fig. 26). In the figure, a solid line represents a cut of the catenoid by a vertical plane passing through the axis of

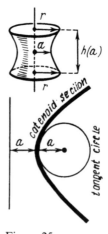

Figure 25

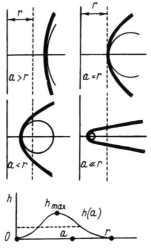

Figure 26

symmetry, while a fine line shows a circle tangent at the catenary curve vertex. For $a = r$, it is obvious that $h = 0$, i.e., the two circles are tangent. When $a < r$, the distance h increases, the circles move apart, and minimal surface sags from them. For some time h keeps on increasing, while a is further decreasing. Then, as seen in Fig. 26, h starts decreasing on attaining a certain maximum value h_{max}. When a tends to zero, h tends to zero, too. The catenary curve approaches the Ox-axis nearer and nearer. The obtained graph of the function $h(a)$ is shown in Fig. 26. When a is close to zero, the curvature radius at the curve vertex is also small, which makes the curve approach the horizontal axis. The maximum value h_{max} is approximately equal to $\frac{4}{5}r$. When the distance a between the circles exceeds h_{max}, the catenoid bursts, and the minimal surface transforms into two flat disks spanning boundary circles. In this case, there are no any other soap films with this boundary. The described passage from one solution to another occurs abruptly, changing the film topology. It is useful to consider the evolution of the catenoid with h increasing from zero to h_{max}. In Fig. 27, you can see several consecutive positions of the soap film. The curious fact is that, for any h, $0 < h < h_{max}$, there are two soap films-catenoids: exterior and interior. When h increases, these two films begin sagging in the direction of the axis of symmetry, drawing near each other. Finally, when h exceeds the critical value, both films burst and change into a pair of disks spanning the circles. Thus, for any value of h, contained in the interval $0 < h < h_{max}$, there exist three soap films with given boundary, i.e., two catenoids and a pair of disks. Plateau suggested that the interior catenoid should be unstable in contrast to the exterior, but he could not prove it. Attempts to realize the interior catenoid in the form of a physical soap film fail. The assumption of instability of the interior film holds.

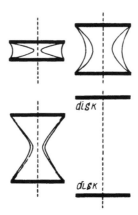

Figure 27

10. The Change of the Topological Type of Minimal Surfaces Depending on Their Stability or Instability

Generally speaking, a few different minimal surfaces can span one and the same frame. We are already familiar with some such cases, e.g., two catenoids and a pair of disks. The uniqueness of a soap film can be guaranteed, e.g., for flat closed non-self-intersecting frames, i.e., realizing the form of a system of curves embedded in one plane. It is clear that any film different from the part bounded by these curves has greater area than the flat film. Meanwhile, two variational problems can be considered: Looking for films of smallest area for a given frame, and minimal films of smallest topological genus. Nevertheless, it is not unlikely that the topological genus of the film with smallest possible area is not minimal; the vice versa, a minimal film of smallest genus cannot realize area minimum in the class of all films with given boundary. In other words, the absolute minimum of the areal functional can be attained on the films of non-minimal topological genus (Fig. 28). If the represented pairs of circles are sufficiently near, then the area of the first film is smaller than that of the second. The first film is homeomorphic to the punctured torus in Fig. 28, i.e., in this sense, it possesses genus 1, while the second film is homeomorphic to a disk and has genus zero. Therefore, for the time being, we dwell on minimal films meaning their local minimality and posing no questions about their global minimality and the relation of absolute minima of the areal functional to the topological genus of the film.

Consider the frame in Fig. 28 which turns out to be sufficiently convenient to observe non-trivial surgeries of a topological type, depending on the stability or instability of soap films. It is clear that the frame is homeomorphic to a circle. We assume that it is realized in R^3 in the form of a closed wire frame. Let

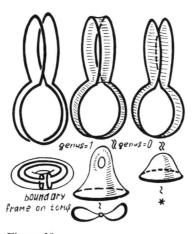

Figure 28

u and v be distances between the upper and lower rings (Fig. 28). For each fixed state of the frame, generally speaking, there exist several types of minimal surfaces. Our aim is to examine their surgeries under frame deformation. Since a formal analytic study of minimal surfaces on the whole is difficult, we had to resort to a physical experiment. The results of the analysis conducted by the author together with A. A. Tuzhilin are listed below. During the experiments the following soap films were obtained: (a), (b), (c), (d), (e) (Fig. 29–31), and (f), (g), (h) (Fig. 32). The films (b). (c), (d), and (e) are diffeomorphic to the disk. Films of type (f) are not manifolds any more, and they contract by themselves to a point (they are homotopy trivial) as well as the films of types (b), (c), (d), and (e). For the present, we confine ourselves to the study of (b),

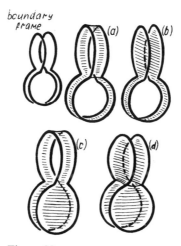

Figure 29

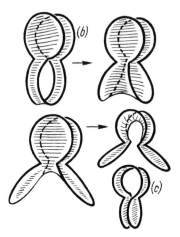

Figure 30

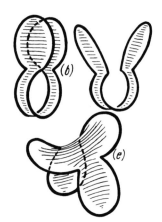

Figure 31

(c), and (d), i.e., films diffeomorphic to the disk. T. Poston in [19] pointed to the film surgery arising under a continuous frame deformation.

Take (b), and deform the boundary frame, causing deformation of its spanning film as is shown in Fig. 30. Gradually move two lower circles apart.

We may assume that this process roughly coincides with the stretching of a catenoid when its two boundaries move from each other. Here, we make use of the fact that when the original distance between the circles is sufficiently small, the soap film, the tape, spanning each pair of circles is sufficiently well approximated by a catenoid. The catenoid being stretched, there comes a critical moment when the catenoid collapses, transforming into a pair of disks spanning two circles moving from each other, whereas in our case these two

disks are still joined with a narrow tape spanning two upper circles which are
less affected by the described deformation. As a result, we obtain the film
shown in Fig. 30. Now, moving the two lower circles together in the reverse
direction and returning them to the original position, we do not change the
topological type of the film any more, and as a result we obtain (c). Thus, we
have constructed a continuous deformation of the frame, resulting in an
abrupt change of the film and its type: from (b), we have passed into (c).

By frame symmetry, we can make a transition in the reverse direction from
(c) to (b) in an analogous way. It is sufficient to perform the above-mentioned
procedure with upper circles, gradually moving them apart from each other
and then returning into the original position. In the process of deformation,
the film undergoes an abrupt surgery and (c) transforms into (b). This
"pendulum" can be realized repeatedly if the soap film is obtained from a fairly
elastic liquid.

In addition to the described abrupt transformation of the film, there exists
another curious boundary frame deformation resulting in a smooth deformation of the film with the same final result (!). Indeed, consider the film (b) and,
in contrast to the previous case, begin moving the upper circles as shown in
Fig. 31. The topological type of the film does not change, it being subject to a
homeomorphism. Then, let us move apart the lower circles as shown in Fig.
31. As a result, we obtain a symmetric film of saddle type, its center being the
saddle point. The film ascends in one direction, and descends in the other. This
deformation is smooth, without abrupt changes. Since the film (e) is
symmetric, we can pass from it to the film (c) in a smooth way, applying the
same operation. In an analogous way, repeating all steps in inverse order, we
can smoothly transform the film (c) into the film (b) without jumps and
topological surgeries. Thus, in the space of all surfaces with homeomorphic
boundary frames of the above type, we obtain two essentially different paths
joining the position (b) to (c). The first path was realized by another smooth
deformation, during which the film underwent a jumplike surgery. The second
was realized by another smooth deformation, during which the film transforms without jumps, passing from (b) into (c) as in the first case. To a first
approximation, we can assume that the above deformations of frames can be
described by change in two real parameters: u, the distance between the upper
circles, and v, the distance between the lower circles. In other words, u and v
are the width of the upper and lower rings-tapes on the surface (a) (Fig. 29).
We can assume that each position of the boundary frame is determined by
specifying a pair of real numbers for visuality assumed to be the coordinates of
a point in the two-dimensional plane.

T. Poston in [19] put forward a method to describe this process by means of
a so-called *cusp*. We decided to tackle this question from another standpoint,
i.e., study the properties of the real graph of soap film area. It appears that the
resulting picture is described by a swallowtail. Sharpen the problem. For each
point (u, v), there is the corresponding frame, generally speaking, spanning

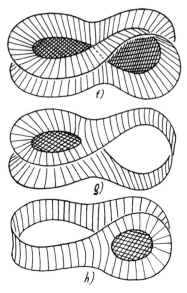

Figure 32

several films. Computing their areas, we obtain the graph of multi-valued function specifying the areas of minimal surfaces with given boundary. There are two types of films: stable and unstable. Among the above films, (d) is unstable while the rest are stable. Let us concentrate on unstable films. We rely on one of Plateau's principles (see below), according to which part of a minimal film bounded by an arbitrary closed frame spanning this film is minimal, too, with respect to this frame. In real experiments, unstable films were obtained by the author and A. A. Tuzhilin in the following way. The instability of a film means that there are its arbitrarily small perturbations destroying the film or causing an essential change in its position relative to the boundary frame. At the same time, if sufficiently many threads with ends attached to the frame are placed conveniently on the film, film stability can be attained provided the thread network is fairly dense. In other words, we can try keeping an unstable soap film in a stable state by increasing the boundary frame; e.g., take the unstable film (d) (Fig. 29) spanning the frame with a thread a priori attached to it as shown in Fig. 33(a). Let this thread lie freely on the film, i.e., no forces act on it. Deform the frame slightly. Moving the right circles apart, we find out that part of the films curves to the left, and the thread stretches, turning into a singular edge of the film, preserving its equilibrium and hampering further deformation. The like occurs if the left circles are moved apart (here, the thread curves to the right and stretches). In the intermediate state, i.e., on the film (d), the thread does not stretch. If in this critical state of the film the thread is detached from one end (instead of

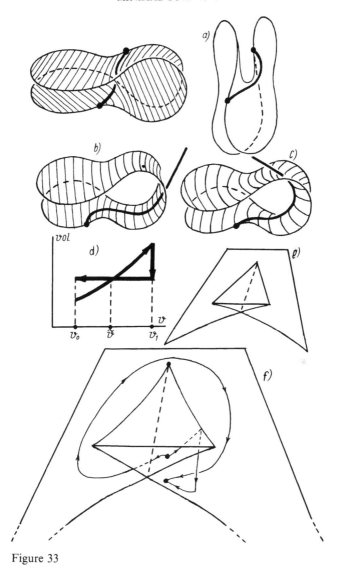

Figure 33

attaching the second end of the thread to the frame, we could hold it in the hand) by being given a slight initial impulse, the film will jump into a state (b). To show that in each state of the frame (for small u and v) there exists an unstable film of type (c), take the film (b), attach a thread to one point of the frame (Fig. 33(b)), and begin pulling the thread (Fig. 33(c)). The film soon finds itself in critical position. If the process is continued, an abrupt change in the film occurs, and the thread curves to the left. If in the critical state of the film the thread is loosened by being given a slight impulse, the film regains its status quo ante (b). Consider now the state of our system $(u_0, v_0(b))$, where u_0 and v_0 are small, and increase v smoothly. The film (b) curvs more to the center, and, finally, such a state sets in when, v being further slightly increased, the film begins moving independently under the action of surface tension forces, and jumps into state (c), the state of the frame being hardly changed. If we now decrease v to the original state, there occur no jumps of the film. It preserves its type (c) resulted from the jumps. For want of space we skip certain details and give only the final graph of the areal change of the film under the above deformations (Fig. 33(d)). The value $\tilde{v}$ corresponds to the state of the frame in which the area of (b) is equal to the area of (c).

The result of our analysis is shown in Fig. 33(e); see the details in [71], [69] (A. T. Fomenko and A. A. Tuzhilin). Drawn is the graph of a multi-valued function describing the area of minimal surfaces of the above type. The obtained surface (and its corresponding singularity) is called a "swallowtail". It is known that any smooth mapping of a three-dimensional manifold into a three-dimensional manifold is approximated by the mappings having Whitney singularities only: the fold, the cusp, and the *swallowtail*.In particular, the swallowtail may be represented as a surface in a three-dimensional space of polynomials of the form $x^4 + ax^2 + bx + c$, composed of points (a, b, c) corresponding to polynomials with multiple roots. We now return to the areal graph of soap films. In Fig. 33(f), it is easy to observe an interesting possibility of transition from the state $(u, v, (b))$ into $(u, v, (c))$ both by a jump and in a continuous (smooth) way. The first way is to shift through the singular edge (path to the right in Fig. 33(f)), and then the "fall" from the point on the edge into the point of type (c). The second way: Descend smoothly of point (b) along the graph sheet, pass round the singular point, and arrive at point (c). Thus, unstable fims of type (d) (from the standpoint of their areal graph) fill in the upper triangle of the swallowtail, and the wings correspond to the stable films (b) and (c). On each singular edge, smooth folia of the graph are tangential with an order of contact $3/2$ as in the classical "swallowtail".

11. Two-Dimensional Minimal Surfaces in a Three-Dimensional Space and the First Plateau Principle

In the thirties and forties of this century, great progress was made in the study of propertis of two-dimensional surfaces in a three-dimensinal space. Remarkable results were obtained by Douglas, Radó, Courant and others. An important role in these studies was played by the Dirichlet principle which we formulate in the following form.

Let G be a plane domain bounded by a Jordan curve γ, and g be a function continuous on the closure of G, piecewise smooth in G, and such that the Dirichlet integral $D[g]$ is finite on this function. Consider the class of all functions φ continuous on the closure of G, piecewise smooth in G, and assuming on the boundary of the domain the same values as the function g. Then the problem of finding the function φ on which the Dirichlet integral $D[\varphi]$ assumes the smallest value has the unique solution. This function is also the unique solution of the boundary-value problem for the Laplace equation $\Delta\varphi = 0$ with a priori prescribed boundary values $\bar{g} = g|_\gamma$ on the curve γ. Recall that the expression of the form $D[\varphi] = \frac{1}{2}\iint (E + G)dudv$, where E, G are coefficients of the first fundamental form (Riemannian metric) induced on the graph of the function by the underlying Euclidean metric in $\mathbf{R}^3$. If $r = r(u, v)$ is the radius vector of the surface, then $E = \langle r_u, r_u \rangle$, $G = \langle r_v, r_v \rangle$.

In the first half of our century, a certain change in approaches to solving the Plateau problem took place. Plateau himself had formulated several principles which we are going to list in the sequel. Here, we only give the first principle.

The First Plateau Principle. Let a surface of zero mean curvature be given. The surface can be described by means of an equation, some geometric rule; e.g., a helicoid or in the form of radius vector. Consider an arbitrary piece-wise closed non-self-interesting frame, i.e., we draw a closed curve on the surface. We assume this frame to enclose part of the surface, this part being stable. Let us make an iron wire frame accurately reproducing the drawn curve, oxidize it slightly with a diluted nitric acid, and immerse it into soapy water (with some glycerin added). If we take the frame out of the water, we are sure to find (among soap films which may appear on this frame) a film coinciding exactly with the part of the surface, bounded by the curve drawn a priori.

It is obvious that Plateau proceeded from the already given minimal surface and then considered various frames which allowed to realize some or other stable parts of the surface in the form of soap films. With the development of variational calculus predominant has become the standpoint that the frame is a primary object, and the problem consists in studying different minimal surfaces which can span the frame. The existence of a soap film spanning the given frame is a top priority. Since the study of arbitrary frames is sufficiently non-trivial, first, a fairly special class was distinguished.

Consider surfaces given in $\mathbf{R}^3$ by a smooth graph of a single-valued function $f(x, y) = z$, i.e., admitting a smooth orthogonal bijective projection on a two-dimensional plane or a domain in it. Let a surface M be projected into a domain G with convex boundary γ, i.e., the boundary curve (supposed to be piecewise smooth) is convex. If the surface is given by the graph, the areal functional has the form $A[f] = \iint_G \sqrt{1 + f_x^2 + f_y^2}\, dxdy$. The Euler equation for the areal functional, i.e., the equation describing critical or stationary surfaces is $(1 + f_x^2)f_{yy} - 2f_x f_y f_{xy} + (1 + f_y^2)f_{xx} = 0$. It is non-linear and complicated behaviour of its solutions, i.e., minimal surfaces, depending on the nature of the boundary frame. It turns out that, for any closed piecewise smooth contour γ' projecting onto the plane into a convex and closed curve γ in a one-to-one

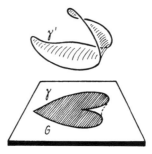

Figure 34

fashion, there exists a minimal surface with given boundary, therefore, projecting onto the domain bounded on the plane by γ.

The existence and uniqueness of this solution in the indicated class of surfaces admitting a one-to-one projection is the consequence of the boundary projection convexity. If we omit the requirement that the domain G should be convex, then the theorem statement becomes invalid. In other words, there exist smooth contours γ' projecting onto the non-convex curve γ in the plane in a one-to-one fashion, and such that there is no minimal surface with boundary γ' projecting onto G bounded by γ in a one-to-one fashion (the simplest example being in Fig. 34). The minimal film spanning γ' does exist; however, first, it is not projected inside γ, and, second, it cannot be given as the smooth graph of a one-valued function defined on the plane. It is clear that these effects are the consequence of γ non-convexity.

12. Area Functional, Dirichlet Functional, Harmonic Mappings, and Conformal Coordinates

Further essential progress has been made by Douglas who considered smooth mappings of the standard two-dimensional disk $D \subset R^2$ into R^3. We refer the disk to Euclidean coordinates u, v. If E, F, G are the coefficients of the first fundamental form of the surface which is the image of the disk in R^3, then we obtain the explicit expressions

$$D[r] = \frac{1}{2} \iint_{D^2} (E + G) \, du \, dv, \qquad A[r] = \iint_{D^2} \sqrt{EG - F^2} \, du \, dv.$$

for the Dirichlet and area functionals [1,5]. The extremal, or critical, radii vectors, i.e., solutions to the Euler equations are as follows. The extremal vectors $r(u, v)$ for the areal functional are those and only those vectors for which the mean curvature of the surface vanishes, i.e., minimal surfaces. The extremal vectors for the Dirichlet functional are those and only those vectors which are harmonic with respect to u, v, i.e. $\Delta r = \dfrac{\partial^2 r}{\partial u^2} + \dfrac{\partial^2 r}{\partial v^2} = 0$. If conformal

coordinates are given on a minimal surface, then its radius vector becomes harmonic. Thus, each minimal film is harmonic with respect to conformal coordinates. The converse does not hold: The surface swept out by the harmonic radius vector must not necessarily be minimal, the simplest example being the graph of the real $a(x, y)$ (or imaginary $b(x, y)$) part of a non-linear complex-analytic function $a(x, y) + ib(x, y)$ given on the plane. The interaction of minimal and harmonic radii vectors is schematically represented in Fig. 35. If some or other radius vector is minimal (i.e., the surface corresponding to it is minimal), then all the other radii vectors obtained from the original by a regular change of variables are also minimal. In these changes the surface remains unaltered, and mean curvature is an invariant of coordinates change. In each of such classes of equivalent minimal radii vectors (with respect to variavle changes), there must necessarily be one harmonic. It is shown in Fig. 35 that each equivalence class intersects the set of harmonic vectors.

The area and Dirichlet functionals are related by the inequality $D(r) \geqslant A(r)$, equality occurring if and only if $E = G$, $F = 0$, i.e., where coordinates u, v are conformal, which follows from the obvious inequality $\dfrac{E + G}{2} \geqslant \sqrt{EG - F^2}$.

Some analytic advantage of the Dirichlet functional over the areal functional is in its integrand not containing any radicals. A similar situation occurs in the theory of one-dimensional length and action functionals, where the extremals for the action functional $\int \langle \dot{\gamma}, \dot{\gamma} \rangle \, dt$ are geodesics and those of the length functional $\int \sqrt{\langle \dot{\gamma}, \dot{\gamma} \rangle} \, dt$ are all possible regular parametrizations of geodesics [2]. In the one-dimensinal case, the arc length functinal has "more" extremals than the action functional. In the two-dimensional case, the picture is more complicated. The most important result proved by Douglas is the existence theorem for a minimal surface in the class of surfaces of fixed topological type. Before stating it, we recall that each smooth, compact, closed and connected two-dimensional manifold is diffeomorphic either to a sphere with a number of handles or a sphere with a number of cross-caps glued (Fig.

Figure 35

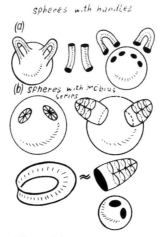

Figure 36

36). A handle is the usual cylinder glued with its two boundaries to two holes in the two-dimensional sphere. A cross-cap is the usual Möbius strip glued with its boundary (along the circle) to the hole boundary in the sphere. Sometimes, the Möbius strip is for visuality realized as a cross-cap (Fig. 36), for which its boundary should be bent, so that it may turn into the standard circle, with the consequence that the Möbius strip configuration gets more complex, and self-intersections may appear. We can now discuss the Douglas result. Assume that, for a two-dimensional manifold M of the given topological type (i.e., with a given number of handles or Möbius strips) with boundary, there exists a smooth mapping $f : M \to R^3$ with finite area, i.e., $A[f] < \infty$, the surface boundary being mapped onto the given set of closed Jordan curves in R^3. Assume that the minimum (more precisely, the infimum) of the areas of such surface immersions into R^3 with fixed boundary is strictly less than the minimum of the areas of all immersions into R^3 (with the same boundary) of all surfaces obtained from the original by discarding one handle or Möbius cross-can. Then there exists a generalized conformal mapping $f_0 : M \to R^3$ which is minimal, i.e., it realizes the area minimum in the class of all surface immersions into R^3 with given boundary. This surface has zero mean curvature at all its non-singular points. We comment on the condition for at least one immersion of M into R^3 with finite area to exist. A similar condition was also used in stating the Dirichlet principle. As a matter of fact, there exist curves which cannot span a surface with finite area (Fig. 37). Two orthogonal projections of a space curve onto two planes are represented in Fig. 37. The curve equation in standard spherical coordinates ρ, θ, φ is written as $\rho = \cos \theta$, $\varphi = \tan^5 \theta$, where ρ is radius vector length, φ is the polar angle on the (x, y)-plane, and $\dfrac{\pi}{2} - \theta$ is the angle made by the radius vector with the vertical

Figure 37

z-axis. As θ tends to $\pm \pi/2$, $|\varphi| \to \infty$, $\rho \to 0$, which gives the representation of two projections. Direct computation shows that this curve is not a boundary of any surface with finite area, immersed into R^3. Moreover, there are examples of such exotic curves whose each point has a similar complicated structure, therefore, spanning any arbitrarily small part of the boundary requires that a surface of infinite area should be constructed. We dwell on the properties of minimal surfaces whose existence is stated by the Douglas theorem.

Since we consider smooth mappings of a surface in R^3, which are immersions at all points except, possibly, a set of measure zero, minimal films may have self-intersections and even branch pints. It is clear that there exist contours for which there is no minimal film in the embedding class; e.g., the simplest knotted contour (trifolium) in Fig. 38, spanned in R^3 by no embedded disk. At the same time, there exists a smooth mapping of the disk D^2 in R^3 (with self-intersections), realizing the minimum of area in the class of

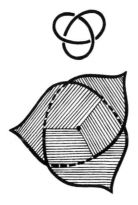

Figure 38

mappings with given boundary. This statement for disk mappings, by the way, holds for any Jordan closed contour homeomorphic to the circle. An example of a minimal film spanning a circle knotted as a trifolium, and immersed in R^3, is also represented. This surface has a branch point (in the center) along with self-intersections (three concurring line segments). However, this surface is unstable.

Many works have recently appeared dedicated to the study of indices of different variational functionals; particularly, of volume functionals and the Dirichlet functional [91], [115], [116]. In this section, we examine indices of the area functional of two-dimensional minimal surfaces in R^3 and H^3 or, simply, indices of minimal surfaces. There exist at least four types of indices which we are going to describe. In [119], R. Bohme and A. Tromba consider generalized minimal surfaces of the type of a two-dimensional disk D^2 in R^n. They prove that the set of these minimal surfaces generates a manifold stratified by the number and type of branch points (singularities) on a minimal surface; the projection setting each minimal surface in a correspondence with the frame it spans is a Fredholm mapping (tangential map at each point is a Fredholm linear mapping). Bohme and Tromba computed the index of this Fredholm mapping (on each stratum of a manifold of minimal surfaces). It appears that the index depends only on the number and types of branch points as well as n, the dimension of the ambient space R^n; e.g., knowing this index enables us to establish that only a finite number of minimal disks can span a typical frame boundary in R^n, and what kind of singularities may be encountered on typical minimal surfaces in R^n.

Three remaining types of indices are varieties of the index of Morse type for the case of compact and non-compact minimal submanifolds. The index of a compact minimal submanifold with boundary in a Riemannian manifold is determined as the index of the second variation of the volume functional (index of a symmetric bilinear form) on the space of normal variations of a submanifold, fixed on the boundary [23]. This index generalizes the classical index of geodesics, and many results pertaining to geodesics hold for it. In particular, the Simons theorem [23] (particular case of the Smale theorem [117]) generalizes the theorem on the relation of the index of geodesics and the multiplicity of conjugate points for the case of a compact, connected and minimal submanifold with boundary. This index is responsible for the stability of a minimal submanifold: If the index is different from zero, then there exists a deformation decreasing the volume of the submanifold. Two-dimensional minimal surfaces in R^3 are a mathematical model of soap films. If a soap film spanning a wire frame is stable, then it realizes a local minimum of the area functional, which causes the index to vanish; i.e., mathematical stability can be spoken of [71], [69].

Many works are dedicated to the study of stability. Barbosa and Do Carmo showed [120] that an immersed minimal surface in R^3, having the image area of a Gaussian map less than 2π, is stable. Another example of checking the index for the stability of a minimal surface is the result of Lawson and Hsiang [26], and A. T. Fomenko and A. V. Tyrin [27] (equivariant stable minimal cones of codimension one in R^n). Here, they succeeded in estimating the indices of minimal cones with the help of the indices of geodesics on the quotient space.

For a non-compact minimal submanifold, we define two types of indices: smooth and piecewise smooth. An interest in the study of indices of non-compact minimal surfaces is determined by the basic examples of minimal surfaces in R^n being complete minimal surfaces which are non-compact.

DEFINITION 1. The supremum of the indices of compact subdomains with smooth boundaries in a non-compact connected minimum submanifold of a Reimannian manifold (by compact domain, we mean a domain with compact closure) is called the *smooth index* of the Riemannian manifold.

DEFINITION 1'. The supremum of the indices of compact subdomains with piecewise smooth boundaries of a minimal submanifold from Definition 1 is called the *piecewise smooth index*.

It is apparent that a smooth index does not exceed the piecewise smooth index. The problem of their coincidence is open at present.

In his works, Do Carmo studied stability in the sense of piecewise smooth index. Thus, in [121], he proved that a complete stable immersed minimal surface in R^3 can only be a plane.

In this section, we concentrate on smooth indices of non-compact minimal submanifolds. The theorem of Smale and Simons is naturally generalized to the class of the smooth index of a non-compact minimal submanifold of a Riemannian manifold. It would be useful to generalize this theorem to the case of a piecewise smooth index, and find out whether the smooth and piecewise smooth indices coincide. As a matter of fact, it is often easier to make computations for piecewise smooth frames than for smooth. Piecewise smooth are, e.g., the "coordinare" contours $\{|u| \leqslant a, v = \pm b\} \cup \{u = \pm a, |v| \leqslant b\}$, where (u, v) are coordinates on the surface, and a, b are two constants. Such frames were used in the works of A. Yu. Borisovich [114], [123].

In [27], A. T. Fomenko has computed a lower estimate for the volume of non-contractible globally-minimal submanifolds whose index is zero. It would be interesting to compute a lower estimate of the volume of a minimal submanifold with a fixed (larger than zero) index.

In [124], an attempt was made to compute indices of classical minimal surfaces in R^3, which failed, since the author instead of the Jacobi equation on the surface where the Laplacian is metric applies the Jacobi equation for a minimal surface, written in conformal coordinates, where a Euclidean Laplacian is used.

The results listed are due to A. A. Tuzhilin. Here and in the sequel, we assume the index to be smooth without specifying it each time.

THEOREM 1 (due to A. A. Tuzhilin). *Let M be a non-compact two-dimensional immersed orientable minimal surface in* R^3 *given globally by the* **Weierstrass representation** (see the definition below)

(a) $(f(w), (aw+b)^m)$, $a \neq 0$ by C, where m is a positive integer;

(b) $(f(w), (aw+b)^m)$, $a \neq 0$ by $C \backslash \left\{ -\dfrac{b}{a} \right\}$, where m is an integer different from zero;

(c) (f, g) on C, where g is an integral function which is not a polynomial. *Then, in* (a) *and* (b), *the index of M is* $2|m| - 1$; *in* (c), *infinite.*

If the domain of Weierstrass representation is restricted, then the index does not increase.

In Theorems 2 and 3, we consider some restrictions placed on the Weierstrass representation U domain, which guarantee that the index of the minimal surface given by this representation over U either does not change (compared with the case of U being the same on C or $C \backslash \left\{ -\dfrac{b}{a} \right\}$ from Theorem 1) or it is zero. These restrictions are given in terms of a Gauss map.

Define a closed subset of the sphere $K \subset S^2 = \{x^2 + y^2 + z^2 = 1\}$ at one of the following points:

(i) For $m > 0$, assume that $K = S^2 \cap \left\{ z \leqslant \dfrac{m-1}{m} \right\}$ if $m = 1$, take as K a closed hemisphere which does not contain the North pole $N = (0, 0, 1)$;

(ii) For $m \in Z \backslash \{0\}$, let K be part of the sphere S^2 bounded between two parallel non-coincident planes which are at the distance $\tanh t_{m-1}$ from the center, the North (and, therefore, the South) pole not being contained in S^2. Here, t_{m-1} is the unique positive solution of the equation $\dfrac{m-1}{m} = \tanh t \tanh \left(\dfrac{m-1}{m} t \right)$.

THEOREM 2 (due to A. A. Tuzhilin). *Let M (just as in Theorem 1) be given by the Weierstrass representation* $(f(w), (aw+b)^m)$ *over* U, *where m is an integer different from zero. Then the index of M does not exceed* $2|m| - 1$. *If U contains the inverse image under the Gausse map of the subset K of the sphere S^2, defined above in* (a) *or* (b), *then the index of M is equal to* $2|m| - 1$.

Define an open subset K' of the sphere S^2 at one of the following points:

(a') $K' = S^2 \cap \{z < 0\}$; if $m = 1$, then, as K', take an open hemisphere which does not contain the North pole;

(b') K' is part of the sphere S^2 bounded between two parallel non-coincident planes at the distance $\tanh t_0$ form the center of the sphere, the North pole not being contained in S^2. Here, t_0 is the unique positive solutin of the equation $t \tanh t = 1$.

THEOREM 3 (due to A. A. Tuzhilin). *Let M (just as in Theorem 1) be given by the Weierstrass representation $(f(w), (aw+b)^m)$ over U, where m is an integer different from zero. If U is contained in the inverse image under the Gause map of the subset K' of the sphere S^2, defined above in (a') or (b'), then the index of M is zero.*

NOTE. The function g in the Weirstrass representation (f, g) has a simple geometric meaning, viz., it is equal to the composition of the Gauss map and stereographic projection. This enables us to write down restrictions placed on the domain U from Theorems 2 and 3 directly, without resorting to the Gauss map. The condition for immersion of the minimal surface M in our case is equivalent to f vanishing nowhere.

The indices of many classical minimal surfaces in R^3 are computed immediately by means of Theorems 1, 2 and 3.

COROLLARY. For the folloing classical minimal surfaces, we give the corresponding Weierstrass representations, and determine their indices:

the *Enneper surface*, $(1, w)$ on C, the index is 1;

the *catenoid*, $\left(-\dfrac{1}{w^2}, w \right)$ on $C \backslash \{0\}$, the index is 1;

the *Richmond surface*, $\left(w^2, \dfrac{1}{w^2} \right)$ on $C \backslash \{0\}$, the index is 3;

the surface *helicoid*, $(e^{-w}, -ie^w)$ on C, the index is ∞;

the *Scherk surface* over a fundamental domain, $\left(\dfrac{1}{1-w^4}, w \right)$, $U = \{|w| < 1\}$,

the index is zero;

the complete Scherk surface, theindexis ∞;

the complete *Schwartz-Riemann surface*, the index is ∞;

the two-fold covering over *Henneberg surface* without branch points,

$\left(2\left(1 - \dfrac{1}{w^4} \right), w \right)$ on $C \backslash \{0, 1, -1, i, -i\}$,

the index is not greater than 1.

Thus, the indices of fundamental classical minimal surfaces in R^3 have been computed. The indices of the surface of the catenoid and Enneper surface were computed independently by Fischer-Colbrie in [128].

We now give the results of investigating the indices of minimal surfaces in the three-dimensional Lobachevski space H^3. Let $L^{n+1} = R_1^{n+1}$ be a pseudo-Euclidean space with the metric dS^2 of signature $(1, n-1)$, and let H^n be realized as a connected component of a unit sphere in the metric dS^2 with the induced Riemannian metric $dl^2 = -dS^2|H^n$.

In [122], the surface of revolution in H^n is defined. Let P be a two-dimensional plane in L^{n+1}. Three cases are possible: P is Lorentz (if the metric dS^2 restricted on P is non-degenerate and alternating); Riemannian ($dS^2|P$ is positive defined and non-degenerate); $dS^2|P$ is degenerate. Let Π be a three-dimensional subspace in L^{n+1}, so that Π contains P and $\Pi \cap H^n$ is non-empty (therefore, $\Pi \cap H^n$ is the Lobachevski plane). In $\Pi \cap H^n$, consider a curve C which does not intersect P. It is well known that all motions of H^n are the restrictions of pseudo-orthogonal transformations in L^{n+1}, carrying H^n into itself. Let G_p denote a subgroup of the group of motions of H^n, leaving all points from P fixed. Then the orbit of C under the action of G_p is called a *hypersurface of revolution* of C about P. A hypersurface of revolution which is minimal is called a catenoid. There are three types of catenoids: spherical (P is Lorentz), hyperbolic (P is Riemann), and parabolic (P is degenerate).

Consider the case $n = 3$. It is proved in [122] that hyperbolic and parabolic catenoids are stable. In [125], Morrey considered a spherical catenoid, and showed that all spherical catenoids generate a one-parameter family M_b, $b > \frac{1}{2}$; if $b > \frac{17}{2}$, then M_b is stable. In [122], a necessary stability condition of a complete immersed surface with a finite total curvature is found, and it is shown that, for $\frac{1}{2} < b < C_0$ where $C_0 \simeq 0.69$ spherical catenoids M_b are unstable.

A. A. Tuzhilin has computed the index of a spherical catenoid, and he has given a simple generic stability criterion for a spherical catenoid.

THEOREM 4. *The index of unstable spherical catenoid in the Lobachevski space H^3 is unity. The spherical catenoid M_{b0} is stable if and only if a variation field of the catenoid M_b in the class of catenoids M_b is nowhere tangent to M_{b0}.*

Then let us introduce cylindrical coordinates into H^3

$t = \cosh z \cosh r$
$x_1 = \sinh r \sinh \varphi$
$x_2 = \sinh r \sinh \varphi$
$x_3 = \sinh z \cosh r$, $H^3 = \{t^2 - x_1^2 - x_2^2 - x_3^2 = 1, t > 0\} \subset R_1^4(t, x_1, x_2, x_3)$

A surface given by the equation $\varphi = az$, $a \neq 0$, is called a helicoid. It is easily verified that the helicoid thus defined in H^3 is a minimal surface.

The following is an easy consequence of the explicit form of the Jacobi equation.

THEOREM 5. *Let M be a two-dimensional orientable minimal surface in the Lobachevski space H^3. Let the Gaussian curvature K of the surface M satisfy the condition $|K| \leq 1$. Then the index of M is zero.*

COROLLARY. *The index of a helicoid $\{\varphi = az\}$ in the Lobachevski space H^3 with $|a| \leq 1$, $a \neq 0$, is equal to zero.*

NOTE 3. *It is interesting to find out what occurs in the case $|a| > 1$.*

Let us turn to the details. First of all, define the Weierstrass representation.

DEFINITION 1. Let U be a domain on the complex plane C. A pair of complex-valued functions (f, g) on U is called a *Weierstrass representation* if f is holomorphic, g is meromorphic, and fg^2 holomorphic. Define the forms $\varphi_1 = -\frac{1}{2}f(1 - g^2)dw$, $\varphi_2 = \frac{i}{2}f(1 + g^2)dw$, $\varphi_3 = fgdw$ holomorphic in U, where w is a complex coordinate in C. If $\{x^k\}$ are standard coordinates in R^3, then we define a mapping U into $R: x^k = \text{Re} \int \varphi_k$, where the integral is taken along the paths in U, emanating from a certain fixed point.

Important Note. We suppose everywhere that x^k are one-valued functions over U.

If $f \not\equiv 0$ in U, then we obtain an immersed and a minimal surface except isolated points called branch points. The condition that the point $q \in U$ is critical is equivalent to $f(q) \neq 0$ and $fg^2(q) \neq 0$. Such surfaces are called *generalized minimal surfaces*. Any minimal surface in R^3 can be given locally by means of a Weierstrass representation. If the minimal surface is simply connected, then a global Weierstrass representation exists [127]. It is easily seen that the coordinate functions $x^k(u, v)$ where $u + iv = w$ are harmonic over U, and the coordinates (u, v) over C are conformal (since $\Sigma \varphi_k^2 = 0$). It can be shown [127] that if M is given by a Weierstrass representation (f, g) over U and $dS^2|M = \lambda^2(du^2 + dv^2)$, then $\lambda^2 = \left[\dfrac{|f|(1 + |g|^2)}{z} \right]^2$, and Gaussian curvature $K = -\left[\dfrac{4|g'|}{|f|(1 + |g|^2)^2} \right]^2$, where $g' = \partial g/\partial w$.

Now, we give the result relative to the form of the Jacobi equation for minimal surfaces in a space of constant curvature and, in particular, for minimal surfaces in R^3 given by the Weierstrass representation [118], [126].

PROPOSITION 1. *Let M be an immersed two-dimensional orientable minimal surface in a three-dimensional Riemannian manifold W of constant sectinal curvature C, and let a normal bundle $NM \to M$ be trivial. Let $V = T \cdot n$ be a section of $NM \to M$, where n is a field of unit normals, and T, a function. Then, the Jacobi equation has the form $\Delta T - 2(K - C)T = 0$. Here, K is a Gaussian curvature M, Δ is the Laplacian on M.*

In particular, if $W = R^3$, we have $\Delta T - 2KT = 0$.

If $W = H^3$ is the standard three-dimensional Lobachevski space ($c = -1$), then the Jacobi equation has the form $\Delta T - 2(K + 1)T = 0$.

If (u, v) are conformal coordinates on M such that the induced metrix $\overline{dS^2}|M = \lambda^2(du^2 + dv^2)$, then the Jacobi equation is rewritten

$$\frac{\partial^2 T}{\partial u^2} + \frac{\partial^2 T}{\partial v^2} - 2(K - C)\lambda^2 T = 0.$$

In particular, if $W = R^3$ and M is given by the Weierstrass representation (f, g), then we obtain

$$\frac{\partial^2 T}{\partial u^2} + \frac{\partial^2 T}{\partial v^2} + \frac{8|g'|^2}{(1 + |g|^2)^2} T = 0, g' = \frac{\partial g}{\partial w}.$$

Before giving a generalization of the Smale and Simons theorem for the case of non-compact submanifolds, we define the notion of exhaustion.

DEFINITION 2. Let M be a non-compact connected manifold. A family M_t, $t \in R$, of compact subdomains M with smooth boundaries such that $M_t \subsetneq M_s$ under $t < \bar{\bar{S}}$ and for any compact submanifold $K \subset M$ there exists such t that $K \subset M_t$ is said to be the exhaustion of M. It is clear that $\cup M_t = M$. We call the exhaustion smooth if ∂M_t depends smoothly on t.

PROPOSITION 2 (generalization of the Smale and Simons theorem).

Let M be a non-compact orientable connected immersed minimal submanifold in the Riemannian manifold. Let M_t be a smooth exhaustion, so that the index of M_t is equal to zero when t are sufficiently small. Then the index of M is equal to the multiplicity sum of conjugate boundaries. Recall that the boundary of a minimal open submanifold is called conjugate if there exists a non-zero Jacobi field (solution of the Jacobi equation) vanishing on the boundary: The dimension of the vector space of Jacobian fields vanishing on the boundary is called the *multiplicity of the conjugate boundary*.

To prove Theorems 1, 2 and 3, we need two results more. As mentioned in [118] and [120], the Jacobi equation under the Gauss map of a minimal surface in R^3 reduces into the standard equation on the sphere, viz., ito $\Delta_{S^2} T + 2T = 0$, where Δ_{S^2} is a metric Laplacian on S^2 in the standard induced metric. Therefore, the index of a minimal surface in R^3 is fully determined by its Gauss map.

PROPOSITION 3. *Let M_1 and M_2 be two minimal connected immersed orientable surfaces in R^3. Let M_1 and M_2 be isomorphic in the following sense: There exists a diffeomorphism $F : M_1 \to M_2$ concordant with the Gauss maps n^1 and n^2, i.e. the diagram below is commutative:*

$$
\begin{array}{ccc}
M_1 & \xrightarrow{F} & M_2 \\
\downarrow{n^1} & & \downarrow{n^2} \\
S^2 & \xrightarrow{\Phi} & S^2
\end{array}
$$

where Φ is a certain isometry S^2 with the metric induced by R^3. Then M_1 and M_2 have the same index.

To prove point (c) of Theorem 1, we need the following result.

PROPOSITION 4 [128].

Let M be a complete two-dimensional immersed connected minimal surface in R^3. Then the finiteness of the index of M is equivalent to finite total curvature of M.

The sketch of a proof of Theorems 1, 2 and 3 is due to A. A. Tuzhilin. Since isomorphic minimal surfaces have the same indices, it suffices to choose a certain model example of a minimal surface and make computations for it. As a model example for proving point (b) of Theorem 1, we take a m-sheeted winding of a catenoid on itself, i.e., an immersed minimal surface in R^3 given in the following way:

$$(z, \Psi) \longrightarrow \begin{cases} x = \cosh z \cos m\Psi \\ y = \cosh z \sin m\Psi \\ z = z \end{cases}$$

Here, (x, y, z) are the standard coordinates in R^3, and (z, Ψ) are coordinates on the cylinder $R \times S^1$. For $m = 1$, we have the standard catenoid. The Jacobi equation has the form

$$\frac{\partial^2 T}{\partial z^2} + \frac{1}{m^2} \frac{\partial^2 T}{\partial \Psi^2} + \frac{2}{\cosh^2 z} T = 0,$$

For each z, expand T into a Fourier series of Ψ: $T = \dfrac{a_0(z)}{2} + \sum\limits_{k=1}^{\infty} a_k(z)\cos k\Psi + b_k(z)\sin k\Psi$. Since T is a smooth function, a_k and b_k are smooth functions of z. Substituting T in the Jacobi equation, we obtain simultaneous ordinary differential equations whose solutions are written out explicitly [123]. Consider the exhaustion $M_t = \{|z| < t\}$; the index of the minimal surface over M_t is known to be zero for small t [23]. From the explicit form of the solutions, we determine such t that $T(t, \Psi) = T(-t, \Psi) = 0$ for a certain $t > 0$. We obtain the following solutions: A one-dimensional solution space with generator $T_0(z, \Psi) = z \tanh z - 1$ on Kt_0, where t_0 is the unique positive solution of the equation $z \tanh z = 1$; for $k < m$, two-dimensional spaces with a basis

$$T_k^1(z, \Psi) = \left[\left(\tanh z - \frac{k}{m} \right) e^{(k/m)z} - \left(\tanh z + \frac{k}{m} \right) e^{-k/m} \right] \cos k\Psi$$

$$T_k^2(z, \Psi) = \left[\left(\tanh z - \frac{k}{m} \right) e^{(k/m)z} - \left(\tanh z + \frac{k}{m} \right) e^{-k/m} \right] \cos k\Psi \quad \text{on } K_{t_k},$$

where t_k is the unique positive solution of the equation $\tanh z \tanh \dfrac{k}{m} z = \dfrac{k}{m}$. spanning K_{t_0}, where $t_0 = 1$;

for $k < m$, two-dimensional spaces $T_k^1(u, v) = \left(\dfrac{1 - (u^2 + v^2)^m}{1 + (u^2 + v^2)^m} + \dfrac{k}{m} \right) \operatorname{Re}(u + iv)^k$

$$T_k^2(u, v) = \left(\frac{1 - (u^2 + v^2)^m}{1 + (u^2 + v^2)^m} + \frac{k}{m} \right) \operatorname{Re}(u + iv)^k$$

spanning K_{t_k}, where t_k is the unique positive solution of the equation

$\dfrac{1-\rho^{2m}}{1+\rho^{2m}} = -\dfrac{k}{m}$. It is clear that the functions T_0 and T_k^j are smooth in 0, which proves point (a) of Theorem 1.

To prove point (c), we notice first that the minimal surface given by the Weierstrass representation $(1,g)$ on C is complete, i.e., any curve extending to infinity is infinite in length. It follows from the explicit form of the metric represented by functions of the Weierstrass representation

$$dS^2/M = \lambda^2(du^2 + dv^2), \lambda^2 = \dfrac{(1+|g|^2)^2}{4}.$$ Furthermore, the Jacobi equation only

depends on the function g, as follows from Proposition 1; therefore, the index of a minimal surface given in point (c) is equal to the index of the complete minimal surface with Weierstrass representation $(1,g)$ on C. Since g is not a polynomial, it has an essential singularity at infinity; therefore, g assumes all values an infinite number of times. It follows from [127] that if the complete minimal surface in R^3 has finite total curvature, then the Gauss map assumes each value a finite number of times; therefore, the minimal surface with the Weierstrass representation $(1,g)$ on C has infinite total curvature. From Proposition 4, it follows that its index is equal to infinity. This completes the proof of Theorem 1.

Theorems 2 and 3 are obtained if the monotonicity of the index is used: The index of a minimal surface defined over a subdomain of the domain of definition does not exceed the index of the minimal surface defined on the whole domain [117].

The idea of the proof of Theorem 4 consists in the fact that on the catenoid in H^3 such conformal coordinates can be chosen that the Jacobi equation is written with respect to them as $\Delta T + UT = 0$, where Δ is the Euclidean Laplacian, the potential U majorized by the potential of the Jacobi equation for the catenoid in R^3 of similar form $\Delta T + U_{\text{Euc}} T = 0$. This enables us to prove that the index of the catenoid in H^3 does not exceed unity, i.e., the index of the catenoid in R^3. From computations, it follows that one and only one Jacobi field vanishing on the boundary of a certain compact subdomain from the standard exhaustion analogous to the exhaustion of the catenoid in R^3 is the Jacobi field which is invariant relative to the rotation of the catenoid about its axis of rotation and symmetric relative to the unique plane of symmetry of the catenoid, which does not pass through the axis of rotation. All such fields are obtained by means of projection onto the normal bundle of the catenoid of variation fields in the catenoid class. If follows from the symmetries that, for a variation field, one point is sufficient, at which the field there touches the catenoid for the normal component (a Jacobi field; see [23]) to vanish on the boundary of a certain subdomain of the standard exhaustion. If such a field is there, the catenoid index is unity.

To prove Theorem 5, we notice that, from the condition $|K| \leqslant 1$, it follows that the potential in the Jacobi equation (see Proposition 1) is non-positive everywhere. Let $V = T \cdot n$ be a Jacobi field vanishing on the boundary of a certain compact domain with smooth boundary (denote this domain by K). Then $\int_K \Delta T \cdot T = -\int_K UT^2 \geqslant 0$. On the other hand, $\int_K \Delta^T \cdot T = -\int_K \nabla T \cdot \nabla T < 0$, since $T \not\equiv 0$. Therefore, such a Jacobi field does not exist; hence, the index of the minimal surface is zero. Theorem 5 is thus proved. Corollary is obtained after straightforward computations. The infinity of the index for complete Scherk and Schwarz-Riemannian surfaces follows immediately from [128].

13. Singular Points of Minimal Surfaces and Three Plateau Principles

The important property of generic minimal surfaces is that they often have *singular points*, i.e., such that the surfaces are arranged in their neighborhoods in a more complex way than the usual two-dimensional plane disk of small radius. That these singularities are there is the "typical situation" in the sense that, for an arbitrary contour, the soap film spanning it "almost certainly" has singularities. The sets of singular points can be complicated and interesting enough. Quite often, singular points are not isolated, and fill while line segments abutting on several film sheets. What is their nature? Plateau himself experimentally discovered four circumstances which we call the First, Second, Third, and Fourth Plateau Principles.

Second Plateau Principle. Minimal surfaces (of zero mean curvature) and soap bubble systems (of constant mean positive curvature) consist of fragments of smooth surfaces converging along smooth arcs.

Third Plateau Principle. Pieces of smooth surfaces, of which films of constant mean curvature consist, can abut on each other only in one of the following two ways: (a) Three smooth sheets converge on one smooth curve, and (b) six smooth sheets (along with four singular curves) converge in one point, i.e., a vertex.

Fourth Plateau Principle. In the case where three sheets abut each other on a common arc or an edge, they make equal angles of $120°$. If four singular curves (and six sheets) converge in one vertex, these edges make equal angles of about $109°$.

The corresponding geometric constructions are in Fig. 39. Each of the represented soap films is realized as a surface spanning the corresponding contour, because the films are stable. It is convenient to imagine one of the mechanisms for forming three surface sheets converging on a common edge at the angle of $120°$. Consider the contour inside a big soap bubble (Fig. 40). Letting the air leave the bubble gradually, we see it starting to envelop the contour and, finally, collapse onto the surface of required form. We will say that a point of a minimal surface is of multiplicity k if k surface sheets converge at it. A regular point is one of multiplicity two: Two sheets converge along a common edge at the angle of $180°$, i.e., the surface is smooth in the

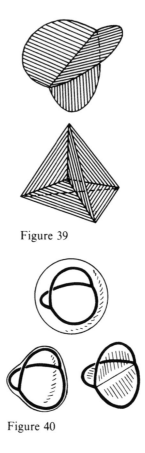

Figure 39

Figure 40

neighborhood of the point. It can be seen in Fig. 41 that two sheets of a minimal surface cannot converge on a common edge at an angle other than π; otherwise, the film can be deformed with a decrease in its area. If we consider surfaces with singular points, then, from the mathematical standpoint, there exist films whose mean curvature vanishes almost everywhere (at non-singular points), and singular point multiplicity may be greater than three; e.g., two orthogonal planes meeting in a straight line do realize such a surface. Here, the multiplicity of each self-intersection point is four (Fig. 41). Nevertheless, attempts to construct a real soap film with more than three sheets converging on a singular line segment fail. It turns out that any surface of zero curvature with self-intersections is unstable; in particular, it cannot be realized as a stable soap film. This statement is of local nature, and not even of global: An arbitrarily small neighborhood of each intersection point is now unstable. In fact, assume that four sheets converge in some self-intersection line segment. We may suppose that they are pairwise orthogonal; otherwise, the film instability is obvious. Cut out part of the film near the self-intersection line

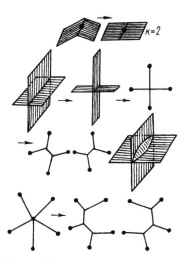

Figure 41

segment, making use of the First Plateau Principle, and fix the boundary of the two thin film strips obtained. Consider the section of the film by a plane orthogonal to the singular segment. The four-fold singular point decomposes into the union of two triple points, since the length of the curve originally consisting of two orthogonal segments is greater than that of each of the curves in Fig. 41 (verify!). Minimizing the length of the curve which is a section by a plane, we minimize the area of the surface piece distinguished. The procedure is represented in R^3 in the immediate vicinity to its section in the diagram. Thus, we have constructed a film deformation decreasing its area. Since the film is now in stable equilibrium (in particular, its small perturbations only increase the area), the resultant of all forces acting at each point of the film is zero. This can be applied also to the singular point. In equilibrium, precisely three sheets converge at each of such points; therefore, it is clear that they meet at the angle $2\pi/3$. Similar argument is applicable in the case where more than four sheets converge on a common edge. The surface gets restructured again occupying a position with lesser area, and then a singular edge of higher multiplicity decomposes into several triple edges (Fig. 41). By the way, we note again that, generally speaking, several soap films can span one boundary frame. It can be seen in the diagram that singular points of large multiplicity can decompose into the union of triple points in several ways.

Above we considered a section of a surface by a plane orthogonal to a singular edge. The deformation constructed was uniform at all points of the edge, near to the section, i.e., this deformation shifts the boundary of orthogonal film strips (Fig. 41). However, it is easy to construct a deformation decreasing film area, also provided the film part boundary represented in Fig. 41 is fixed rigidly. The four-fold edge decomposes into the union of two triple

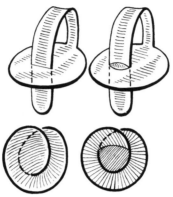

Figure 42

bent edges with fixed ends. Thus, we have constructed a monotonic decrease in the film, decreasing its area for a fixed boundary. Gluing this film piece into the original film of large size, we obtain the required decreasing deformation. The same procedure can be demonstrated if a two-dimensional disk is immersed smoothly into R^3 (Fig. 42). Rising above, the thin strip of the disk intersects the film plane orthogonally. An attempt to realize the surface as a real soap film immediately reveals its instability. The four-fold self-intersection edge decomposes into the union of two bent triple edges with common ends.

These problems are related to the two-dimensional Steiner problem of searching for a thread of least length, joining a finitely many points on the plane. Meanwhile, the thread can be branched, and it can possess points of multiplicity three or higher. The problem can be solved experimentally by means of soap films. Two nearby planes made of thin acrylic plastic should be taken and joined with vertical columns-segments placed at those points of the plane through which the required thread should pass. Dipping the structure into soapy water and removing it, we obtain a film spanning the columns and realizing its plane section length minimum. Meanwhile, the thread contains points of multiplicity not greater than three.

Similar mechanisms act also in the cases where air pressure is essential in forming a soap film; e.g., bubbles filled with air are formed. The film then occupies the position making up for the pressure inside. In the general case, combining the film spanning the contour with the bubbles, we can obtain real films including both bubbles and surfaces of zero mean curvature. This effect is seen, e.g., in foaming.

In Fig. 43, sufficiently complicated boundary frames and some spanning minimal films are represented [11]. The interaction of different surface sheets is seen to be of intricate nature, and it gets more complicated as the boundary frame does. Sometimes, the soap films with bubbles inside bounding closed air volumes form spontaneously when sufficiently complicated frames are removed from the liquid.

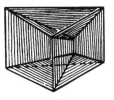

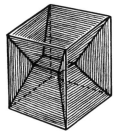

Figure 43

Let us return to the problem of constructing a one-dimensional minimal thread (with possible branching of multiplicity 3) connecting fixed points in the plane. We can imagine trying to pull a rubber thread of least length on nails driven into a plane. It is clear that this problem is connected with the so-called transshipment problems: How to plan a highway system connecting given cities, so that the total highway length would be the least? Theoretically, this problem has not been solved yet for the case of the plane. Intrinsically, the problem makes sense for other surfaces different from the plane, too. Formulate the following problem which is naturally called the *one-dimensional Plateau problem*. Let a two-dimensional (or multidimensional) Riemann surface with certain fixed Riemannian metric (metric is naturally assumed to be sufficiently simple and homogeneous) be given. On the surface, consider a closed one-dimensional net (thread with branching) satisfying these requirements: (i) This net is stable, i.e., its each sufficiently small part realizes minimum length, the ends of this part being fixed. This net is said to be *minimal*. Problem: Classify all minimal nets whose length is equal to the absolute minimum of lengths (in the class of all possible nets of given topological type). In the problem formulated above, the net (thread) has no ends. Certainly, we can consider the version of the problem when the thread is attached to fixed points (i.e., it is given with boundary conditions). It follows from the stability condition that at each branch point of the net, precisely three edges should converge at equal angles.

If a manifold is the standard two-dimensional sphere, then the one-dimensional Plateau problem (without boundary conditions) is solved easily;

all locally stable (but not globally minimal) one-dimensional nets are described below in Sec. 2, 5. Such nets are precisely ten.

A two-dimensional manifold on which it would be extremely interesting to describe all solutions of the one-dimensional Plateau problem is the two-dimensional plane torus, i.e., the torus with locally Euclidean metric. I. V. Shklyanko (Ptitzina) has made some progress in solving this problem. Results obtained by her are given here in brief. Consider a closed (i.e., without end points) one-dimensinal minimal stable net on the plane torus T^2. Then the net decomposes the torus into polygons with angles of $2\pi/3$ at the vertices. It turns out that *all polygons are hexagons, the number of knots (i.e., singularities) of the net is even, and the number of polygons is twice less the number of knots.*

A net is called *regular* if it is composed of congruent hexagons. Then, two nets on the torus are called *equivalent* if there exists a homeomorphism of a net onto the other, carrying the sides of polygons into parallel sides and preserving oriented normals to them.

The statement holds: *On a flat torus, any minimal stable net is equivalent to a regular net.*

From this theorem, the corollary is immediate: *Equivalent nets have equal length.* Since we consider nets on the torus, stable minimal nets are of finite length, the torus being a compact manifold.

It turns out that, on a flat two-dimensional torus, a minimal stable net with 2n knots (i.e., with 2n vertices which are singularities of the net) always exists for any fixed natural number n. This result follows not only from the constructions of I.V. Shklyanko but from the theorem of A. Edmonds, J. Ewing, and R. Kulkarni (Annals of Math., 116, (1982), pp. 113–132). It appears that the number of different (non-equivalent) stable minimal nets with 2n knots can be found, and the obtained result geometrically interpreted. We shall describe this theorem in brief. Assume that the flat torus T^2 is given by a non-singular matrix $A = \begin{pmatrix} a & b \\ c & d \end{pmatrix}$, $a, b, c, d \in R$, i.e., $T^2 = R^2 Z / \oplus Z$, where the action of the group $Z \oplus Z$ on R^2 is given by the matrix A.

Consider two natural numbers k_1, k_2, and two integers m_1, m_2 (they may be positive or negative). These four numbers uniquely specify a net on the torus but we omit the details here. We notice that k_1, k_2, m_1, m_2 have geometric interpretation in terms of a minimal net.

To formulate a theorem conveniently, we introduce two two-dimensional vectors $e_1 = 1/n(ak_2 + bm_2, ck_2 + dm_2)$ and $e_2 = 1/n(-am_1 + bk_1, -cm_1 + dk_1)$. The angle between the vectors α and β is denoted by $\varphi(\alpha, \beta)$. Consider now the system of inequalities:

$$-\tfrac{1}{2} \langle \cos \varphi(e_1, e_2) \langle 1,$$

(1)
$$-\tfrac{1}{2} \langle \cos \varphi(e_1, e_2 - e_1) \langle 1,$$

$$-\tfrac{1}{2} \langle \cos \varphi(e_2, e_1 - e_2) \langle 1,$$

and $\qquad\qquad\qquad\qquad\qquad\qquad\qquad\qquad\qquad\qquad\qquad\qquad$ (*)

$$-\tfrac{1}{2}\langle\cos\varphi(e_2,e_1)\langle 1,$$

(2) $\qquad\qquad\qquad -\tfrac{1}{2}\langle\cos\varphi(e_1,-e_1-e_2)\langle 1,$

$$-\tfrac{1}{2}\langle\cos\varphi(e_2,e_1+e_2)\langle 1.$$

Fix a natural number n, and consider all possible methods for representing it in the form $n=|k_1 k_2 + m_1 m_2|$, where the quadruple (k_1, k_2, m_1, m_2) satisfies the conditions: (i) k_1, k_2 are natural numbers; (ii) m_1, m_2 are integers (i.e., they can be both positive or negative); (iii) the quadruple (k_1, k_2, m_1, m_2) satisfies either only precisely one or both conditions (1) and (2) of the system (*).

Compute now the natural number $p(n)$ by the following rule. We do not distinguish (k_1, k_2, m_1, m_2) and $(|m_2|, |m_1|, -(\operatorname{sgn} m_2)k_2, -(\operatorname{sgn} m_1)k_1)$; we call these quadruples equivalent. If any quadruple satisfies exactly one of the conditions (1), (2) of the system (*), we count this quadruple once, i.e., with multiplicity 1. If a certain quadruple satisfies both conditions (1) and (2) of the system (*), we count this quadruple twice, i.e., with multiplicity 2. Now compute the number of obtained quadruples with multiplicities for the fixed number n (i.e., the total number of all non-equivalent quadruples, each quadruple being taken with multiplicity either 1 or 2; see above). The obtained number is denoted by $p(n)$.

THEOREM (I. V. Shklyanko). *Let a flat (i.e., locally Euclidean) two-dimensional torus T^2 with matrix $\begin{pmatrix} a & b \\ c & d \end{pmatrix}$ be given, where $a, b, c, d \in R$, and the matrix is unimodular. Then the number of different, i.e., non-equivalent nets (this number is equivalent to the number of regular nets on the torus) with $2n$ nodes is equal to the number $p(n)$ defined above.*

We emphasize that in each class of non-equivalent nets on the torus there exists precisely one regular net, therefore, $p(n)$ is a number of regular nets with $2n$ nodes. Incidentally, two nets on the torus may be isometric though not equivalent in the sense of the above definition. For example, certain nets can be rotated so that a new net is obtained on the same torus it is non-equivalent to the original [72].

Then I. V. Shklyanko noted that, among nets with $2n$ nodes on the torus T^2 (where n is any fixed natural number), absolute length minimum is attained on equivalence classes for which the following equalities hold (here, e_1 and e_2 are the same vectors as above):

$$\begin{cases} \cos\varphi(e_1,e_2)=\tfrac{1}{2} \\ \cos\varphi(-e_1,e_2-e_1)=\tfrac{1}{2} \\ \cos\varphi(-e_2,e_1-e_2)=\tfrac{1}{2} \end{cases} \quad \text{or} \quad \begin{cases} \cos\varphi(e_2,e_1)=\tfrac{1}{2} \\ \cos\varphi(-e_1,e_1-e_2)=\tfrac{1}{2}. \\ \cos\varphi(-e_2,e_1+e_2)=\tfrac{1}{2} \end{cases}$$

From here it follows, in particular, that absolute minimal nets with $2n$ nodes on the torus are the nets for which polygons of regular nets equivalent to them are regular.

I. V. Shklyanko carried out in analogous research for the case of two-dimensional closed Riemann surfaces M_g^2 of an arbitrary genus g (i.e., for two-dimensional spheres with an arbitrary number of handles), the standard Riemannian metric of constant negative curvature (equal to -1) being considered on the manifold M_g^2 (of genus $g > 1$). In other words, M_g^2 is considered to be the quotient manifold of the Lobachevski plane with the standard Lobachevski metric. Here, the picture is more complex than in the case of a sphere and torus. We omit the details and concentrate on the facts which are easy to formulate.

THEOREM. *The number of singularities (nodes) of a locally minimal stable net on the surface M_g^2 is always equal to $2n$; the number of polygons $k = 2 - 2g + n$, where g is the genus M_g^2; the polygons have no less than seven vertices (angles). Let l_m be the number of m-gons in the net, then $\sum_{m \geqslant 7} m l_m = 6n$, $\sum_{m \geqslant 7} l_m = 2 - 2g + n$. A closed local minimal stable net with $2, 4, \ldots, 2(2g - 2)$ does not exist on M_g^2. For minimal nets with $2n$ knots on M_g^2 where $g \geqslant 2$, whose polygons are isometric m-gons in the Lobachevski metric the absolute minimum of length (nets) is reached on the nets of which all polygons are regular.*

A classification of all absolutely minimal nets from congruent polygons on M_g^2 up to isometry follows from the A. Edmonds, J. Ewing, and R. Kulkarni theorem.

14. Realization of Minimal Surfaces in Nature

As early as the 18th century, it was known that many concrete problems arising in physics, chemistry or biology could be reduced to the analysis of minimal surfaces. We illustrate by a biological example. It turns out that minimal films are often to be found in nature as most economical surfaces forming the skeletons of certain living things; e.g., Radiolaria, tiny sea animals of most various and exotic form.

The first who paid attention to the essential role of surface tension in shaping these living things was apparently D. A. Thompson, the author of the book *On Growth and Form*. Radiolarians consist of small blobs of protoplasm inside foamlike forms similar to soap films or bubbles. Since these organisms are complicated enough, minimal films involved have, generally speaking, many branch points and triple singular edges on which the fluid of the organism concentrates mostly. The fluid is placed along the singular edges, and can be well displayed on real soap films. It flows off the film freely until the edge where the film sheets converge is encountered. The fluid then is stopped, and forms water segments visually revealing the minimal surface singular edges. The same process, in fact, occurs in Radiolaria. The fluid concentration along the branching edges leads to the sediment if hard fractions of sea water and salts along the edges, which is just what makes the animal hard skeleton. The geometry can be visually represented if the singular edges along which different bubbles are placed inside soap bubbles are scrutinized. These segments are joined sometimes in quite a fantastic manner, and form the "fluid skeleton". After death, soft tissues vanish gradually, and the hard skeleton formed in the manner described above remains. It is nothing but the visual, now permanent, representation (or photograph) of the branching edges and singular points arising in the complex minimal surface which also included closed volumes. Besides, in forming the solid skeleton of a Radiolaria, the cell walls accumulating sea salt are also important. Three Radiolaris skeletons are represented in Figs. 44, 45 (see Ernst Haeckel's Report on the Scientific Results of the Voyage of the HMS Challenger during the Years 1873–1876).

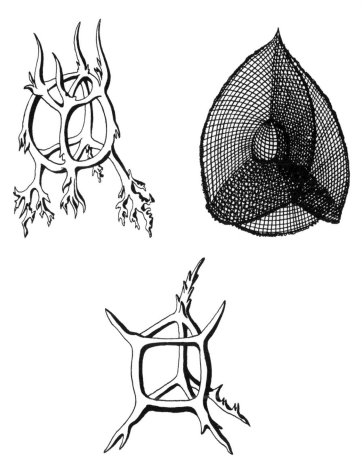

Figure 44

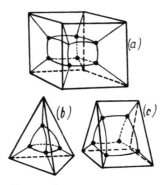

Figure 45

Here, three minimal surfaces with soap bubbles spanning the frames are represented, which are the system of the edges of the usual tetrahedron, cube or prism. The similarity of the form of real organisms with the above films is striking. It can be seen that Radiolari skeletons reproduce branching edges of the minimal surface fairly accurately.

This is not a unique example of the role which the minimal surfaces play in the environment. The films are also important in chemistry where surface interactions on the boundary between different media determine the nature and rate of many chemical changes. The well-known membranes found both in various natural formations and some spheres of human activity can also be taken as examples of minimal surfaces. Such are the eardrum, the membranes between living cells, etc. The main property of a membrane is to strive for a position optimal from the standpoint of energy in the given environment. Therefore, the form of a membrane is determined to a considerable extent by external environment; vice versa, knowing the shape of a minimal surface, we can judge the external actions which are optimal for the organism. Real membranes also possess a number of important properties; e.g., selective penetrability or semi-penetrability enabling the cells to exchange ions.

We see that minimal surfaces are also related to a number of interesting mathematical problems which, being posed correctly and investigated, enable us to understand this curious, though outwardly simple, natural phenomenon better,

2. Topological Properties of Minimal Surfaces

1. On Various Approaches to the Concepts of Surface and Boundary

In this section, we shall attempt to describe the development of investigations of the Plateau problem briefly, starting with the sixties when important developments occurred again attracting the specialists' attention to the classical problem of minimal surfaces. We do not claim even to a brief survey of all new directions which are due to the remarkable works of such authors as H. Federer, W. Fleming, F. Almgren, M. Miranda, E. Reifenberg, Ch. Morrey, E. Bombieri, E. Giusti, E. De Georgi, J. Simons, H. Lawson, W. Allard, et al. Sufficiently complete information and bibliography are contained, e.g., in [7–16, 18, 19]; the interested reader is referred to these works. We confine ourselves to the discussion of only new qualitative effects, topological in nature, which enable us to extend our ideas of the properties of minimal surfaces.

In the previous section, we did not draw the reader's attention to the concept of film or boundary, limiting ourselves either to the intuitive idea or consideration of the class of two-dimensional surfaces with boundary, smoothly mapped into R^3. However, the concepts of surface and boundary are not so simple and unambiguous as they seem at first glance. If in the case of a one-dimensional boundary frame we can have to do with the above ideas, for this multidimensional Plateau problem related to the study of the multidimensional volume functional, the statement itself of the above concepts requires a special study. The progress reached since the sixties is due to a considerable extent to a sufficiently supple formulation of the notion of surface and its boundary.

It follows from the physical properties of two-dimensional minimal surfaces that they should not possess "holes" or punctures which do not interact with the film boundary. Otherwise, such a hole expands under the action of surface tension until the hole film or its part collapses onto the boundary frame. The property is easily verified by way of real experiment. We only have to prick the soap film with a sufficiently thick piece of wire or rod. The above property of soap films is taken as the basis of the mathematical definition of a boundary and the property of a film to span the boundary. The corresponding mathematical idealization is certainly a simplification of real situation, since there exist (and are stable) physical soap films formed on a wire boundary frame, which leave some of its portions free. In other words, a soap film can be stable and simultaneously attached not to the hole frame, but only to its part. This is owing to the finite thickness of a real wire, which stabilizes the mathematically impossible situation in some cases (Fig. 46). We will then consider closed contours; however, we should bear in mind (though we are not going to take up the problem) that there exist real soap films spanning non-closed boundary frames (see the diagram). This contour can be non-closed, since it can be rectified into a line segment. Here, the wire should be sufficiently thick compared with the film size. From the mathematical point of view, there exists no minimal surface (reasonably) spanning the segment embedded in R^3 without self-intersection.

The study of critical (minimal) surfaces for the functional of multidimen-

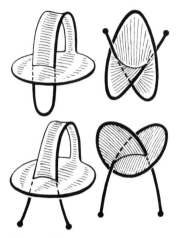

Figure 46

sional Riemannian volume shows that the class of smoothly embedded or immersed submanifolds on which the functional is defined is not wide enough to contain a solution to the variational problem. In other words, a minimal surface may not belong to the class. Accordingly, E. Reifenberg noted that, though it is clear intuitively that any set which is a minimal-area surface should be in a certain sense well arranged locally, this statement should be proved, and the proof cannot be obtained if in the beginning we start considering surfaces only well arranged locally. In other words, it would be of use to investigate the structure of minimal-area sets, not assuming a priori that they resemble manifolds.

2. Homological Boundary of a Surface and the Role of the Coefficient Group

Since we would like to extend the concept of surface, that of its area or volume in the multidimensional case should be extended accordingly. As the fundamental functional of the type defined on closed k-dimensional submanifolds in R^n, we can take the k-dimensional spherical *Hausdorff measure* to be denoted by $\mathrm{vol}_k X$. Its definition and properties can be found, e.g., in [7, 8, 20]. Here, we do not give its formal definition, restricting ourselves to the statement that if a subset X is a smooth submanifold or stratified algebraic submanifold in R^n, then $\mathrm{vol}_k X$ coincides with the usual k-dimensional Riemannian volume of a submanifold. As "surfaces", we can now consider measurable, i.e., of finite Hausdorff measure, compact subsets in R^n. It turns out that this class is now wide enough, and contains minimal surfaces. It remains to formulate the concept of boundary. Fix a homology theory with coefficients in a group G. If a surface X is a finite cell complex or smooth manifold, then we can take the usual cellular or simplicial homology as the q-dimensional homology group $H_q(X, G)$ [1,4]. However, if X is only a measurable compact set, then, as $H_q(X, G)$, we should consider so-called *spectral (Čech) homology*. In the case of finite cell complexes and smooth manifolds, these homology groups coincide with cellular (simplicial) homology groups. Spectral homology is the natural generalization of these last to the case of measurable compacta [7, 8]. Therefore, here we are not going to enter into details of constructing spectral homology, since these particulars are not required, while the interested reader can always reconstruct them, resorting to the standard literature.

It is natural to consider that the surface X (in the wide sense) wholly spans the boundary A if A is contained in X and the surface has no "holes" whose boundary is of the same dimension as A. Homology groups turn out to be adapted enough to formalize this intuitive representatin mathematically. We will say, following J. Adams and E. Reifenberg, that a closed k-dimensional set $X \subset R^n$ is a G-surface spanning a closed $(k-1)$-dimensional set A for a given Abelian coefficient group G if, first, A is contained in X and, second, the homomorphism $i_*: H_{k-1}(A, G) \to H_{k-1}(X, G)$ induced by the embedding $i: A \to X$ is identically zero, i.e., the whole group $H_{k-1}(A, G)$ is mapped into the identity element under addition.

In other words, all $(k-1)$-dimensional cycles contained in the boundary A should "dissolve" in the surface X. In the sense, A is the homology boundary of X.

If we are interested in minimal surfaces in R^3, then A is regarded as the boundary of X if the homomorphism $i_*: H_1(A, G) \to H_1(X, G)$ is trivial. The above definition admits an equivalent reformulation (which is sometimes useful). If we consider a segment of the exact sequence of the pair of spaces (X, A)

$$H_2(X, A, G) \xrightarrow{\partial} H_1(A, G) \xrightarrow{i_*} H_1(X, G)$$

then the condition for the homomorphism i_* to be trivial is equivalent to the condition that the boundary homomorphism $\partial: H_2(X, A, G) \to H_1(A, G)$ is an epimorphism. This definition of the boundary of a surface is to a considerable extent consistant without intuitive idea of a boundary. In fact, consider some particular cases.

(1) Let M^k be a smooth, compact and orientable manifold with boundary A, which is a smooth, orientable $(k-1)$-dimensional manifold. Here, we have the classical situation, or manifold with boundary in the sense of differential topology. Consider the embedding $i: A \to M$. Then the homomorphism $H_{k-1}(A, G) \to H_{k-1}(M, G)$ is trivial and A is the boundary of M in the homology sense.

(2) Let A be an arbitrary finite cell complex, and X the cone over A (Fig. 47).

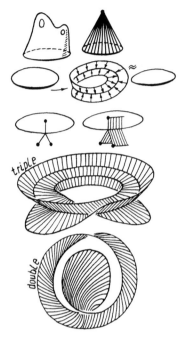

Figure 47

Then the induced homomorphism $H_{k-1}(A, G) \to H_{k-1}(X, G)$ is trivial (for all k), which follows from the whole of the "surface" X contractability on itself to a point.

The concept of G-surface with homology boundary A essentially depends on the choice of the group G. The set A can be the boundary of X for one group, and cannot be the boundary of another. We illustrate by a number of examples.

(3) If A is a circle and X a Möbius strip (non-orientable surface), then the embedding-induced homomorphism $i_* : H_1(A, Z) \to H_1(X, Z)$ has the zero kernel, i.e., is a monomorphism $i_* : Z \to Z$, $i_* : 1 \to 2$, i.e. i_* doubles each cycle, which (from the homotopy standpoint) follows from the equivalence of the embedding $i : A \to X$ to the mapping of A onto itself with degree two (Fig. 47). Thus, a Möbius strip is not a Z-surface with a circle as its boundary, However, a Möbius strip is a Z_2-surface whose boundary is a circle. We obtain for the group $G = Z_2$ that the homomorphism $i_* : H_1(A, Z_2) \to H_1(X, Z_2)$ is trivial, since its reduction modulo two annihilates the image of an integral generator of $H_1(X, Z_2)$.

(4) Let A be a circle, and X a triple Möbius strip obtained by taking an unknotted circle in R^3 and a "trifolium", three line segments of the same length converging in a point of the circle orthogonally to it and making equal angles $2\pi/3$. Shift this figure along the circle, leaving the trifolium orthogonal

to it, but rotating about the axis, so that after a complete rotation, the segments should be permuted thus: $1 \to 2 \to 3 \to 1$ (Fig. 47). The boundary of the surface obtained is a circle traversing about the vertical axis three times. The surface is not a smooth manifold, since it contains as much as the circle of triple singular points. Therefore, a triple Möbius strip is formed according to the same recipe as the usual Möbius strip, but now the boundary traverses the vertical axis three times. If the trifolium were small in diameter, then this surface would be realized as a stable soap film spanning the boundary frame A. However, X is neither a Z-surface nor Z_2-surface with boundary A, which follows from the homotopy equivalence of X and the circle and the fact that the homomorphism $i_* : H_1(A, Z) \to H_1(X, Z)$ sends the generator marked 1 of the group Z into the element marked $3 \in Z$. Therefore, i_* is non-trivial both over Z and Z_2. However, it is trivial for the group Z_3. Therefore, a triple Möbius strip is a Z-surface with the circle as its boundary.

Thus, letting the coefficient group G vary, we can represent different pairs of spaces (X, A) as G-surfaces X with boundary A. It is interesting that the boundary circle A traversing about the axis three times (see Example 3) is at the same time the boundary of the usual Möbius strip (see the diagram). This surface is now without singularities, and also realized by a stable soap film.

If we are not interested in distinguishing the orientable and non-orientable cases, then making use of the group Z_2 is mostly justified in a sufficiently wide class of problems. One of the fundamental results in this direction was the remarkable Reifenberg theory for compact surfaces in R^n, then extended by Ch. Morrey to the case for arbitrary smooth Riemannian manifolds. We formulate it in the general case at once. Let M be a smooth, complete Riemannian manifold, and $A \subset M$ a compact, measurable $(k-1)$-dimensional subset G a compact Abelian group, $H_{k-1}(A, G) \neq 0$, where H_{k-1} are spectral homology groups, and let there exist at least one measurable k-dimensional subset X embedded in M as a G-surface with boundary A. Consider the class $\{X\}$ of all such surfaces, and put $d = \inf\limits_{X \in \{X\}} \mathrm{vol}_k(X \backslash A)$, i.e. d is the infimum of the volumes of all G-surfaces with boundary A. We then assert that $d > 0$ and that there exists a minimal surface X_0 from $\{X\}$, so that $d = \mathrm{vol}_k(X_0 \backslash A)$. Except of the set S of points of k-dimensional measure zero this surface $X_0 \backslash A$ is an open, smooth and minimal submanifold in the sense of differential geometry, i.e., of zero mean curvature. If the ambient manifold M is analytic, then the submanifold $X_0 \backslash (A \cup S)$ is also analytic.

Such surfaces realizing a volume absolute minimum can be naturally said to be globally minimal.

3. Curious Examples of Physical Stable Minimal Surfaces, nevertheless Retracting on their Boundaries

The main restriction on the area of application of this remarkable theorem is the fact that there exist natural soap films spanning a "good" boundary A; however, they are not G-surface with boundary A for any Abelian group G. In other words, the algebraic definition of a boundary does not embrace certain geometric films with the natural physical realization. We illustrate by the interesting example due to J. F. Adams.

We first consider minimal films spanning circles and traversing around a certain axis two or three times in more detail. The boundary frame in Fig. 48(a) clearly spans a soap film diffeomorphic to the usual Möbius strip if the two loops are sufficiently close to each other (Fig. 48(b)). The film is stable and non-orientable, i.e., a Z_2-surface whose boundary is a circle. However, there is another stable field with the same boundary but sufficiently different from the Möbius trip. If the two loops are sufficiently far from each other, then there exists a soap film homeomorphic to the disk and spanning the boundary frame (Fig. 48(c)). If we let two coils approach each other, then the disk may not retain the original position, its parts start shifting towards each other, and, finally, the film may collapse onto a new surface (Fig. 48(d)). This film has a singular edge consisting of triple singular points. A similar phenomenon occurs for a frame making three rotations about the axis (Fig. 48(e)). If the three coils are drawn sufficiently wide apart, then a stable soap film spans the contour, and is homeomorphic to the disk (Fig. 48(f)). It is clear that, as soon as we let the three circle coils approach each other, positioning them as in Fig. 48(e), then a moment comes when the disk gets unstable and collapses onto a new minimal surface now not homeomorphic to the disk (Fig. 48(g)). Note that there exist another two stable non-orientable surfaces with the same boundary as a triple Möbius strip, one of them being the usual Möbius strip (Fig. 48(h)), and the other obtained from the Klein bottle by discarding the

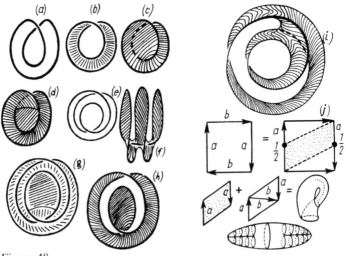

Figure 48

disk (Fig. 48(i)). Recall that the Klein bottle is obtained by gluing two Möbius strips to each other along their common boundary (Fig. 48(j)). It is clear that the surface in Fig. 48(i) is obtained on gluing two Möbius strips (exterior and interior of lesser diameter) together, and discarding the disk.

We now turn to the Adams example. Consider the contour which is the boundary of a triple Möbius strip (Fig. 48(e)).

We already know that there exists a minimal surface with this boundary for the group Z_2 (Fig. 48(g)), where the middle hole formed by the contour is spanned by the disk. The other soap film is a triple Möbius strip (Fig. 47), and a Z_2-surface no more, being a Z_3-surface. It might seem that, selecting each time a convenient coefficient group G, we can represent any "reasonable" film with given boundary as a G-surface. However, this is not so. Consider the boundary frame A in Fig. 49(a) which is the boundary of the stable soap film X in Fig. 49(b) obtained if we glue a double and triple Möbius strips together and join them with a thin bridge. It is obvious that this film is realized by a physical soap film, and quite a "reasonable" surface with given boundary. The boundary of the film is the circle embedded in R^3 without self-intersections.

However, this film is not a G-surface with a circle as its boundary for any Abelian group G. As a matter of fact, the film can be retracted onto its boundary! Recall that a subspace A of a space X is called a retract of X if there exists a continuous mapping $f:X \to A$ defined on the whole of X and leaving all points of A fixed. In other words, the film "instantaneously" collapses onto its boundary which is left fixed. We give the proof below, only giving a necessary corollary. Let $i:A \to X$ be an embedding, and $f:X \to A$ a retraction. The composite mapping $A \xrightarrow{i} X \xrightarrow{f} A$ coincides with the identity. The composite homomorphism

$$H_1(A, G) \xrightarrow{i_*} H_1(X, G) \xrightarrow{f_*} H_1(A, G)$$

is the identity; in particular, i_* is a monomorphism. Therefore, for any group

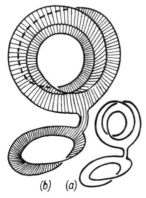

(b)   (a)

Figure 49

G, the film X does not span any cycle in $H_1(A, G)$, i.e., X is not a G-surface with boundary A. Moreover, the homomorphism $i_*: F(A) \to F(X)$ is a monomorphism for any natural "functor" F; e.g., for homotopy groups. Therefore, A is not the boundary of X in any reasonable algebraical sense, though X is a stable film and realizes the absolute minimum of area in the class of all "reasonable" films with boundary A.

We prove that A is the retract of X by making use of the Hopf theorem (Ch. 1, Sec. 2). We take S^1 as A. Let X be the union of a triple and double Möbius strip. Compute the homomorphism $i_*: H_1(A, U) \to H_1(X, U)$. Since the space X is homotopy equivalent to the wedge of two circles, $H_1(X, U) = U \oplus U$ and $i_*(\lambda) = (2\lambda, 3\lambda)$, where $0 \leqslant \lambda \leqslant 1$. It can be seen that i_* is a monomorphism. Q.E.D.

Note that, though A is the retract of X, A is not the deformation retract of X. Recall that a subspace $A \subset X$ is a deformation retract if there exists a continuous homotopy $\varphi_t: X \to X$ such that φ_0 is the identify mapping of X onto itself and φ_1 maps X into A, leaving A fixed for all t. Since the composition $A \xrightarrow{i} X \xrightarrow{\varphi_1} A$ is the identity mapping and the composition $X \xrightarrow{\varphi_1} A \xrightarrow{i} X$ is homotopic to the identity mapping, X and A are homotopy equivalent. In this case, A is a circle, X is equivalent to the wedge of two circles, i.e. A and X are not homotopy equivalent, or A is not a deformation retract.

The existence of a retraction has been obtained on the basis of the algebraic properties of these complexes. It is convenient to have a visual idea how the retraction occurs geometrically. The construction was performed by T. N. Fomenko [67]. We make use of the geometric proof of the Hopf theorem in Ch. 1. Here, the complex X is two-dimensional, and its one-dimensional skeleton coincides with the wedge of two circles, one of which is the usual Möbius strip axis, and the other the triple Möbius strip axis. In the last case, the circle consists of the singular points of a soap film. Proceeding as in Sec. 2, Ch. 1, consider the cellular partition of X. It is clear that $X = \sigma^0 \cup \sigma_1^1 \cup \sigma_2^1 \cup \sigma^2$, where σ^2 is a unique two-dimensional cell. How can it be glued to the one-dimensional skeleton? Cut the soap film along the axes b and a respectively of the double and triple Möbius strips and along the shorter segment c which is the axis of the membrane-bridge joining these two (Fig. 50(a)). The film X is representable as $X = \sigma^0 \cup a \cup b \cup c \cup k^2$, where k^2 is an annulus. On cutting along the one-dimensional skeleton, the film turns into a plane domain homeomorphic to the usual annulus k^2 whose boundary consists of one exterior circle, the original boundary A of X, and one interior circle, the union of seven arcs (three copies of the segments a, two copies of the segment b and two copies of the segment c; see Fig. 50(b)). For convenience, we carry out a deformation retraction when the interior boundary touches its exterior boundary at one point (Fig. 50(c)). The complete boundary obtained of the two-dimensional domain encloses the two-dimensional cell σ^2. To establish the correspondence with the notation in Theorem 3, Sec. 2, Ch. 1,

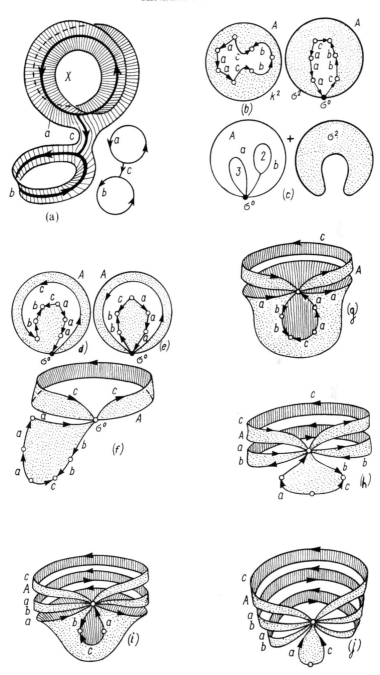

Figure 50

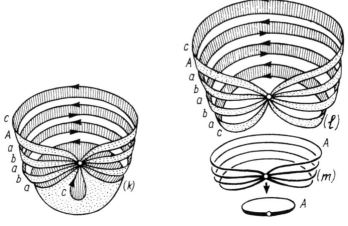

Figure 50 (*cont*)

denote A by S_0^1 (distinguished circle in the one-dimensional skeleton), the circle a as by S_1^1, and the circle b by S_2^1. Then the incidence numbers of the cell σ^2 for these circles are $[\sigma_2 : A] = 1$, $[\sigma^2 : a] = 3$, $[\sigma^2 : b] = 2$. Following the construction in Sec. 2, Ch. 1, we have to construct a mapping g of the one-dimensional skeleton of X onto A. Map A onto itself identically. The most important now is to define g on a and b. We have to select such integers m and n that $m \cdot [\sigma^2 : a] - n \cdot [\sigma^2 : b] = 1$, i.e., $3m - 2n = 1$. We then have to construct a mapping g sending a onto A with degree m, and b onto A with degree n. It suffices to put $m = n = 1$ in our case, i.e., to consider that a and b are mapped onto A homeomorphically. By Theorem 3, Sec. 2, Ch. 1, this mapping g is extendable to the mapping f of the whole film onto A, which is just what is required. It remains to represent this mapping visually. It turns out that, after we have cut the film along the one-dimensional skeleton, the required mapping f is realized in R^3 as a sequence of continuous deformations of σ^2. We first fix A, and extend the side c, so that it might be wound onto A, with degree unity (Fig. 50(d)). Meanwhile, we do not identify A with c, and stop the procedure, preserving the narrow strip bounded in Fig. 50(e) by A and c. The remaining part of the two-dimensional cell forms a disk glued at the point σ^0 in the strip obtained in between A and c. We take the edge a, and wind it along A as shown in Figs. 50(f). The disk hanging off σ^0 turns around A, and we obtain a bag with boundary a, the bag having a hole whose boundary is $bbc^{-1}aa$. The word is written out in counterclockwise motion on the hole boundary, while the degree $+1$ or -1 indicates the orientation of the corresponding edge (Fig. 50(g)). We continue winding the two-dimensional cell. Take b, also winding it along A.

As a result, we obtain the surface represented in Fig. 50(h). The boundary of the hanging disk has become simpler compared with the previous step, and is of the form $a^{-1}a^{-1}cb^{-1}$. Take the next copy of the edge a and wind it along A. Thus, we consecutively alternate the winding of edges of a- and b-type (result is in Fig. 50(i)), and a bag with boundary a and hole bca is obtained again. Take the remaining copy of the edge b (it is now the last), and wind it along A. Finally, we obtain the surface represented in Fig. 50(j), containing the hanging disk with boundary $a^{-1}c$. Take the last copy of the edge a, and repeat the winding. We obtain the surface as in Fig. 50(k) with a bag whose boundary is a and which contains a hole with boundary c. As the last step, we extend c, blow the hole up and turn the bag into a narrow strip enveloping (Fig. 50(l)). The continuous deformation construsted of the cut film X is remarkable, since the film now consists of four narrow strips joined at one point σ^0 and placed in space along the circle A in its immediate vicinity. All the edges a are now in the direction opposite to that of a. The two edges c are also oriented similarly (in the direction of a). Since all edges of one type are oriented similarly, we arrive at the final stage. Contract each of four strips onto its axis, midline, and obtain four circles in Fig. 50(m), mapping them in the same fashion onto A. The constructed mapping is continuous, since all the edges of the same type on the surface in Fig. 50(l) are consistent, which just completes the construction of the required retraction. The same mechanism is also applicable in the case where we consider the union of p-ple and q-ple Möbius strips, where p and q are prime to one another, and not of double and triple Möbius strips. In principle, the visual representation of the retraction can be made use of here, too; however, the winding of the circles a and b onto the circle A, with degrees m and n, is added, where m and n are such that $mp - nq = 1$. This winding makes the geometric picture more complicated but does not alter the principle for constructing the retraction.

We now give the second example due to J. F. Adams, the minimal film X spanning the contour A in Fig. 51(a), homeomorphic to the circle embedded in R^3 without self-intersections. The film represented in Fig. 51(b) is obviously a "good" minimal film, though unstable and with singularities. However, it can be shown not to be a G-surface with boundary A for any group G, since it can be retracted onto its boundary. Moreover, there exists not only the usual retractin, but also a deformation retraction of X onto A. Contracting the film continuously onto its boundary, leaving the boundary fixed, is in Fig. 51 [19], which means that the destruction of the above unstable film occurs gradually, and not abruptly as in the first example: The film continuously deforms on itself, shortens, and, finally, contracts onto its boundary. Two handles on the right and left (Fig. 51(b)) are formed by pairs of coaxial circles attached to the main frame. Since the circles are near to each other, a soap film catenoid-like in shape spans them. The gorge of each is spanned by a plane disk glued on the inside to the circle of lesser radius on the catenoid. We concentrate our attention on the two singular points Q and Q' where four singular edges converge at approximately $109°$. Besides, six film sheets enter each point. Near

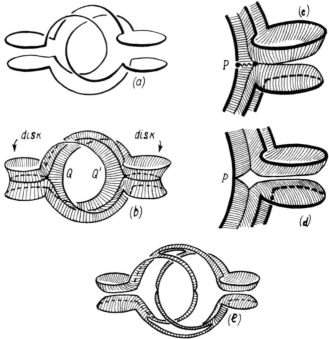

Figure 51

to Q, Q', the film is arranged as in the neighborhood of the tetrahedral film center (Fig. 39). To represent the deformation retraction, it is convenient to make the film slightly thicker, which enables us to treat the sliding of the film on itself more freely. When part of the film shifts on itself, it is convenient to represent these parts separately, making use of the film "thickness". Recall that the deformation occurs inside the thickened film. We first contract on itself to a point each of the plane disks spanning the gorges of the side catenoids (Fig. 51(c)). We can assume that the disks are sucked into Q and Q'. Therefore, two convex disks spanning the upper and lower cricles on the separate handle, e.g., on the right, can be regarded as non-touching inside the thickened film. Hence, the line segments c (Fig. 51(c)) can be contracted to a point P, and the situation represented in Fig. 51(d) occurs. The free space obtained, or a hole, irrevocably starts expanding along the film, decreasing its area monotonically (Fig. 51(e)). A tongue is drawn out of each handle, absorbing its part of the surface in shifting along the former circle of triple singular points. Finally, these tongues return to two pairs of lateral disks, also absorb them, which completes the contraction process. With all the curious properties, it is to the point to recall that we have already encountered no less remarkable (though simpler) examples of real films spanning non-closed boundary frames (Fig. 46).

3*. Realization of the "Bing House" in the Form of a Minimal Surface in R^3

S. V. Matveev and A. T. Fomenko pointed out that the Adams example can be converted into realizing the "Bing house" in the form of a minimal surface in R^3. This hypothesis was proved by O. Yu. Soboleva whose result is given here.

Recall how the "Bing house" is constructed. Take a solid cube in R^3, and bisect it by a horizontal plane (Fig. 52(1). Place a worm on each half of the cube, so that each worm might eat its half through, and make for the "neighboring territory". Inside the adjacent half, the worm eats away the whole interior of the cube, leaving intact the walls, the faces of the cube, the horizontal plane, the wall of the tunnel away by the neighbor, and the membrane joining the walls of the tunnel and the house, We obtain a "house with two rooms", which is a closed two-dimensional cell complex. This complex is fairly well known in topology under the name of the "Bing House". It is remarkable for being homotopy-equivalent to a point though it is not contractible to a point combinatorially. In other words, it cannot be knocked down, rejecting a two-dimensional simplex at each step, since the complex has no boundary; therefore, there is no simplex to begin with (there is no simplex with at least one free face).

Consider sets of singular points in the Bing house and the Adams film shown in Fig. 52(2). It is obvious that these sets are homeomorphic between themselves and homeomorphic to three tangent circles. We call boundary those points of the complex, the neighborhood of which is homeomorphic to a half of the disk. The boundary of the Adams minimal film is a circle. Cut a small disk out from the Bing house. We obtain a two-dimensional complex with boundary, a circle. Stretch a hole on the Bing house, so that the boundary of the complex passes near singular points of the Bing house as shown in Fig. 52(3). A set of singular points on our two complexes (Adams and Bing with a hole) is a union of three circles (Fig. 52(4)) which we denote by X, Y, Z. Fix orientations on the circles composed of singular points and on the boundaries of both complexes. Since each of our two complexes is arranged so that its boundary is near to the set of singular points, a complex may be given in the following way. Moving along the boundary circle of the complex, we write out the letters X, Y, Z one by one. If, in traversing, opposite orientation is induced, the corresponding letter (X, Y or Z) is labeled -1. As a result, we obtain a certain word (composed of the symbols X, Y, Z) which uniquely determines, obviously, our complex up to a homeomorphism; i.e., if the two words built coincide, then the complexes are homeomorphic. Now, write out two words corresponding to the original complexes, the Adams film and the Bing house with a hole. We obtain $XYZ^{-1}Y^{-1}ZYX^{-1}Y^{-1}$ and $XYZ^{-1}Y^{-1}Z^{-1}YXY^{-1}$ (Fig. 52(5)). These words are different; they differ in orientatins of X and Z (in two places). Therefore, the Adams complex can be slightly modified to make the words coincide completely. We try to change the

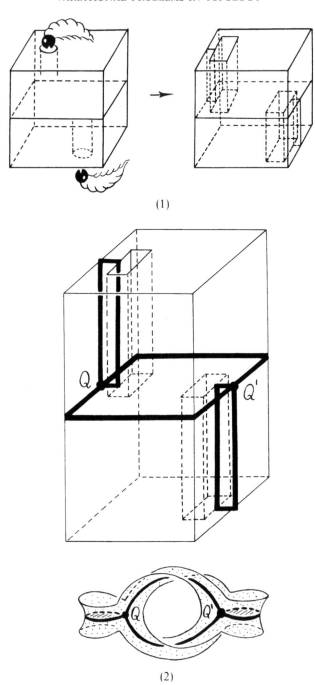

(1)

(2)

Figure 52

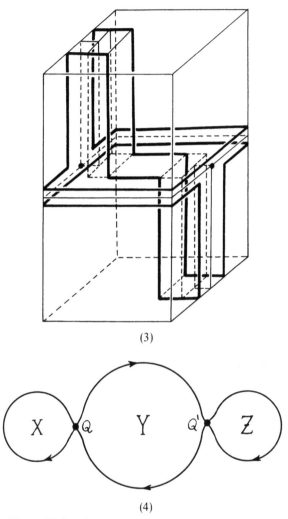

(3)

(4)

Figure 52 (*cont*)

boundary circle embedding in R^3 by deforming the circle smoothly
(Fig. 52(6)). Another minimal film shown in Fig. 52(7) spans this new
boundary circle. The new minimal film may be "broken" into three parts, two
catenoids and the middle "annulus" sewn up from three sheets. In the Adams
example, this annulus is a triple Mobius strip obtained in rotating and shifting
the trifolium along the circle at 120° (Fig. 52(8)). In an example given by
Soboleva, the trifolium does not rotate about its own axis. The new film is also
a deformation retract; it contracts into itself to its boundary circle as shown in
Fig. 52(9). The word is the same as that corresponding to the Bing house with a
hole (see Fig. 52(10)). Thus, the following proposition is proved.

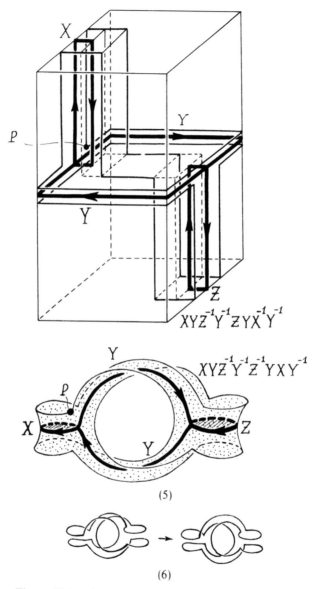

$$XYZ^{-1}Y^{-1}ZYX^{-1}Y^{-1}$$

$$XYZ^{-1}Y^{-1}Z^{-1}YXY^{-1}$$

(5)

(6)

Figure 52 (*cont*)

PROPOSITION (due to O. Yu. Soboleva). *The minimal film shown in Fig. 52(7) is homeomorphic to the Bing house with a hole (i.e. to the Bing house with a disk removed. Therefore, the Bing house with a hole is embeddable into three-dimensional Euclidean space as an absolutely minimal surface (with the boundary circle shown in Fig. 52(6) on the right).*

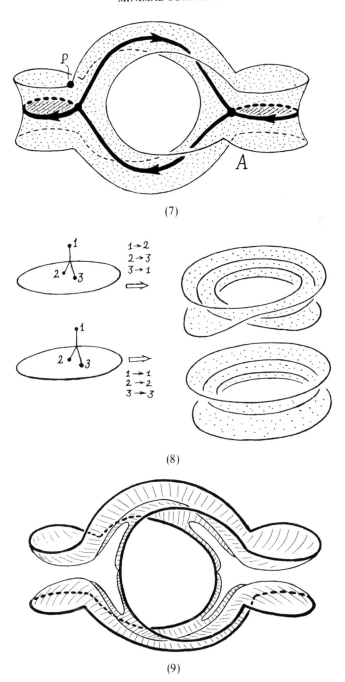

(7)

(8)

(9)

Figure 52 (*cont*)

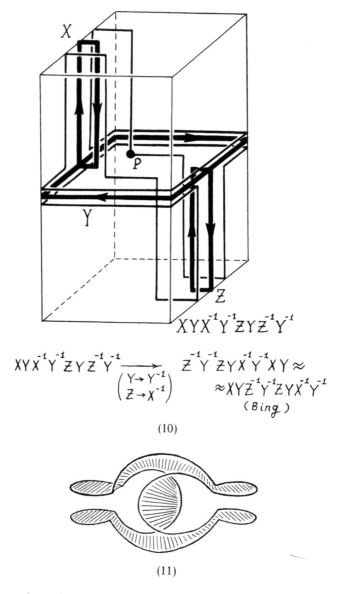

$$XYX^{-1}Y^{-1}ZYZ^{-1}Y^{-1}$$

$$XYX^{-1}Y^{-1}ZYZ^{-1}Y^{-1} \xrightarrow[\left(\begin{array}{c}Y \to Y^{-1}\\ Z \to X^{-1}\end{array}\right)]{} Z^{-1}Y^{-1}ZYX^{-1}Y^{-1}XY \approx$$

$$\approx XYZ^{-1}Y^{-1}ZYX^{-1}Y^{-1}$$

$$(Bing)$$

$$(10)$$

$$(11)$$

The boundary of the new minimal film (Fig. 52(7)) is homeomorphic to the circle. Let a new disk shown in Fig. 52(11) span the boundary circle. Obviously, this disk can be considered to be a minimal surface embedded without self-intersections in R^3. Adding this new disk to the minimal film in Fig. 52(7), we obtain a new minimal film represented in R^3 with self-intersections. It is clear that we have built a map in general position of the Bing house into R^3 in the form of a minimal surface.

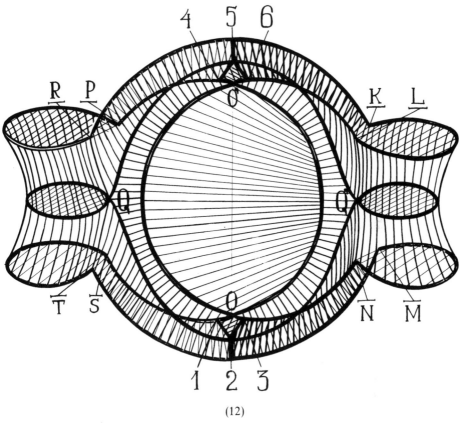

(12)

Figure 52 (*cont*)

PROPOSITION. *The Bing house can be mapped (in general position) onto a minimal surface in three-dimensional Euclidean space. This mapping (in general position) is shown in Fig. 52(12).*

In Fig. 52(12), all basic points and lines of the minimal film are labeled. The classical Bing house in which the corresponding points and lines are labeled is shown in Fig. 52(13). When the disk spans the minimal film shown in Fig. 52(7), self-intersection lines of the film appear. A pair of segments on the Bing house corresponds to each of these singular segments. This correspondence (Figs. 52(12), (13)) is given by the following table.

$$1' + 1'' = 1$$
$$2' + 2'' = 2$$

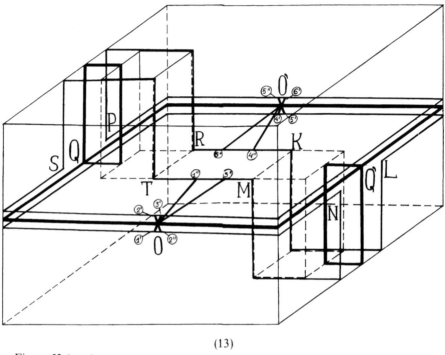

(13)

Figure 52 (*cont*)

$$3' + 3'' = 3$$
$$4' + 4'' = 4$$
$$5' + 5'' = 5$$
$$6' + 6'' = 6$$

Nos. of segments on Bing house	Nos. of segments on minimal film

4. When Does the Soap Film Spanning a Boundary Frame Not Contain Closed Soap Bubbles?

We now turn to the study of global topological properties of minimal surfaces X of dimension k, spanning in the algebraic, or homology, sense $(k-1)$-dimensional "contours" embedded in a certain Riemannian manifold M. We already know that if $H_{k-1}(A, G) \neq 0$ and if there exists at least one film spanning A, then there is always a minimal surface X with boundary A, realizing the absolute minimum of volume (areal) functional in the class of all G surfaces with boundary A. In practical experiments, we often obtain films containing both portions of surfaces of zero mean curvature and soap bubbles containing bounding space. The question arises: (a) when does a soap film consist only of surfaces of zero mean curvature, and (b) when does it contain surfaces of constant positive mean curvature, or bubbles? "Real" minimal films, i.e., wholly containing surfaces of zero curvature, or without bubbles, are of particular interest. There are simple criteria which the boundary A should satisfy for the film realizing the absolute volume (area) minimum not to contain bubbles. These were discovered by the author.

The mathematical concept of homology cycle is an excellent model of a real soap bubble. In some cases, we can imagine a cycle to be a closed, smooth surface, i.e., without boundary. Homology cycles of dimension k naturally form the k-dimensional homology group of space X_0. Therefore, from the topological standpoint, the question whether or not the k-dimensional surface X_0 contains 'bubbles" admits the following restatement: Does X_0 contain k-dimensional cycles (for the given coefficient group)? In other words, k-dimensional surface X_0 contains no "bubbles" if and only if $H_k(X_0, G) = 0$; e.g., all three soap films in Fig. 44 contain closed bubbles and, therefore, possess a non-trivial two-dimensional homology group which is non-trivial and isomorphic to the coefficient group G, i.e., the group $H_2(X, G)$ has one generator. Living Radiolaria associated with these films possess the same property (Fig. 44). On the contrary, the films represented in Figs. 38, 39, 48, 51 have the trivial group of two-dimensional homology. Meanwhile, some of the films are quite non-trivial; in particular, they do not contract on themselves to a point. Thus, to clarify whether or not the minimal surface X_0 contains "bubbles", i.e., cycles, we have to compute $H_k(X_0, G)$.

PROPOSITION 1 (A. T. Fomenko [13, 15]). *Let M be a complete, smooth Riemannian manifold, and $A \subset M$ a compact, measurable $(k-1)$-dimensional subset; e.g., $(k-1)$-dimensional submanifold in M. Suppose that G is an Abelian group, and a one-dimensional vector space over a field F; e.g., let G be the group R^1 of real numbers or $G = Z_p$, where p is a prime number. Let there exist at least one G-surface in M with boundary A. Then the globally minimal G-surface X_0 with boundary A possesses the property that $H_k(X_o, G) = 0$, i.e., does not contain k-dimensional cycles, i.e., "bubbles".*

See the proof below. If the ambient manifold M is arranged in a more complex way then Euclidean space, then it may contain non-trivial k-dimensional cycles itself, or "holes", i.e., the group $H_k(M, G)$ may be other than zero. In applications, the cases where the minimal film spanning a boundary $A \subset M$ are often met, also realizing at the same time certain k-dimensional cycles in M, i.e., enveloping the corresponding holes. Therefore, along with the problem of spanning the boundary A there arises no less natural geometric problem of realizing via minimal surfaces those cycles which are contained by the ambient space M. We now define a realization formally.

Fix in $H_k(M, G)$ some non-trivial subgroup P. We will say that a subset $X \subset M$ realizes P, i.e., its cycles, if P is contained in the image of the group $H_k(X, G)$ under the homomorphism $j_* : H_k(X, G) \to H_k(M, G)$ induced by the embedding $j : X \to M$.

Consider the class $[P]$ of all compact subsets X, where $A \subset X \subset M$ realizing the subgroup P. It turns out that a globally minimal surface X_0 always exists in the class, i.e., so that its volume (area) $\mathrm{vol}_k(X_0 \backslash A)$ is $\inf_{Y \in [P]} \mathrm{vol}_k(Y \backslash A)$, [13, 15].

It is natural to expect due to visual representations that the minimal surface X_0 realizing P does not contain any "superfluous" k-dimensinal cycles, and does contain only those enveloping non-zero cycles from the group $H_k(M, G)$. Meanwhile, X_0 may realize some additional cycles not in P, and not only from P. In other words, the homomorphism $j_* : H_k(X_0, G) \to H_k(M, G)$ must be a monomorphism. No cycle on the film X_0 should not vanish without trace in the manifold M if the film is minimal. It is clear intuitively that if a cycle from X_0 turns out to be extra, i.e., became homologous to zero after embedding the film in the manifold, then the part of the film, corresponding to the cycle (or its support) can be discarded without affecting the realizability of the cycles from P and at the same time decreasing the volume (area) of the film. The procedure is impossible for a minimal film; therefore, the film cannot contain superfluous vanishing cycles. This proposition is justified (proved by the author).

PROPOSITION 2 ([13, 15]). *Let the conditions in Proposition 1 be fulfilled. Then if the minimal surface X_0 realizes a non-trivial subgroup $P \subset H_k(M, G)$. the homomorphism $j_* : H_k(X_0, G) \to H_k(M, G)$ induced by the embedding $j : H_0 \to M$ is a monomorphism, i.e., there are no "superfluous" cycles on the film.*

The existence of globally minimal X_0 follows from [7, 8] and [13–15] in a more general situation. We now turn to the proof of Propositions 1 and 2.

LEMMA 1. *Let X be a k-dimensional compact set in M, Z a compact subset of X, G and Abelian group which is a one-dimensional vector space over a field F, and $H_k(Z, G) = 0$. Assume that $X \backslash Z$ is an open topological k-dimensinal submanifold; let $l \in H_k(X, G)$, $l \neq 0$. Then there exists an open k-dimensional disk $D \subset X \backslash Z$ such that if $i_* : H_k(X \backslash D) \to H_k(X)$ is the homomorphism induced by the embedding, then (a) i_* is a monomorphism (b) Codim $\mathrm{Im}\, i_* = 1$ in the group $H_k(X)$ and (c) $l \notin \mathrm{Im}\, i_*$.* Thus, globally minimal G-surfaces with boundary A do not contain (with the above assumptions) non-trivial k-dimensional cycles. i.e., k-dimensional "bubbles", whereas the globally minimal surfaces realizing non-trivial cycles in the ambient manifold do not contain any vanishing cycles. i.e., each bubble of the minimal film must envelop some non-trivial hole-bubble in the ambient manifold. These statements being natural from the intuitive standpoint and hard to imagine to be invalid in some situations, our intuition turns out to be insufficient here, too; in reality, there exist certain G-surfaces (for convenient groups G) containing non-trivial k-dimensional bubbles-cycles. We illustrate by examples. Take the group U of complex numbers with unique modulus, i.e., $\{e^{i\varphi}\}$, as the group G, and not a one-dimensional group over a field. U is representable as $R^1(\bmod\ 1)$, and homeomorphic to the circle. Consider the class of minimal U-surfaces with boundary A; as it turns out, there are globally minimal U-surfaces with boundary A, also realizing non-trivial cycles in the ambient manifold, i.e., enveloping the k-dimensional holes in M, and not only possessing a non-trivial k-dimensional homology group, i.e., containing bubbles-cycles of dimension k. As M, consider the direct product $M = S^1 \times \mathbf{RP}^{2s}$, where $\mathbf{RP}^{2s}$ is a real projective space. Consider the projective subspace $\mathbf{RP}^{2s-1}$ as the boundary A, embedded in $\mathbf{RP}^{2s}$ standardly, and its $(2s-1)$-dimensional skeleton. We take the space $\mathbf{RP}^{2s}$ as the U-surface X_0. It is clear that $H_{2s-1}(A, U)$ is non-trivial, and isomorphic to the group U. Therefore, under the embedding $A \to X_0$, the whole group $H_{2s-1}(A, U)$ is mapped into the zero element, i.e., X_0 is a U-surface with boundary A. Since X_0 is a cofactor in the direct product $S^1 \times \mathbf{RP}^{2s}$, it is easy to show that X_0 is globally minimal among all U-surfaces with boundary A. On the other hand, $H_{2s}(X_0, U) = H_{2s}(\mathbf{R}P^{2s}, U) = \mathbf{Z}_2$, i.e., the group is non-trivial, and X_0 can be easily seen to realize the non-trivial subgroup Z_2 in the homology group M.

5. Minimal Cones Related to Singular Points of Minimal Surfaces

We now retrace our steps to the problem of physical realization of soap films. When the films may or may not contain closed bubbles-cycles. The question is intimately related to the behavior of soap films near to their singular points. We may assume on the basis of the Plateau principles that, in a sufficiently small neighborhood of each singular point, a minimal surface consists of several smooth parts of a two-dimensional surface, which can be assumed approximately plane. We shall see in the sequel that this property may not hold for minimal surfaces of dimensions higher than two. There are "multidimensional" singular points whose no arbitrarily small neighborhood consists of plane parts. Returning to the two-dimensional case, we consider a sphere of small radius, center at a singular point, and study its intersection with the surface. We can assume that the intersection consists of finitely many smooth arcs contained in the sphere and forming a net on it.

It is easy to see that each arc segment should be a segment of an equator, i.e., the circle obtained in the intersection of the sphere by a plane passing through its center. In fact, since we assume the film to consist of plane parts in the vicinity of a singular point, they converge at the point, and the whole film inside the sphere is a cone, with vertex at a singular point. The "basis" of the cone is the set of smooth arcs forming the sphere net. The soap film enclosed by the sphere is made up of all possible radii emanating from the center of the sphere and reaching the one-dimensional net. Some radii may be singular edges of the film where its several sheets converge. The film is formed by several sectors, or parts of the usual plane disk. The length l of the one-dimensional net and area S of the film bounded by the sphere are related simply, since it is clear that $S = \frac{1}{2} r_0 l$, where r_0 is the radius of the sphere. It follows that the net consists of equator segments, i.e., geodesics on the sphere. If we assume that part of an arc is not a geodesic, than there is a small deformation not making its extremities coincident but decreasing its length. Therefore, there is a contracting deformation also decreasing the area of the two-dimensional film-cone, since a decrease in l leads to that in S. Thus, if several smooth net arcs converge at a point on a sphere, then there are three of them and they form an angle equal to $2\pi/3$ [11].

The question arises: What are those one-dimensional nets obtained by intersecting a soap film with a small sphere whose center is placed at one singular point of the film? The answer is description of film structure locally in the neighborhood of its singular point not on the boundary frame spanned by the film, so that (1) each smooth arc is part of an equator, and (2) only three arcs making an angle $2\pi/3$ may converge at each singular point of the net, or its node, thus enabling us to list completely all possible configurations, or nets. There are ten such configurations (all in Fig. 53). To see that there are no other nets, it suffices to make use of elementary argument from spherical trigonometry.

We now clarify whether or not all these configurations are realizable as the intersections of certain soap films with small sphere centered at a singular point. Most of the above configurations are not seen to be as such. If we consider rectilinear cones over these nets, vertices at the centre of the sphere, then they are such to possess zero mean curvature at all their regular points (Fig. 53). If we neglect a set of measure zero, then they are minimal surfaces.

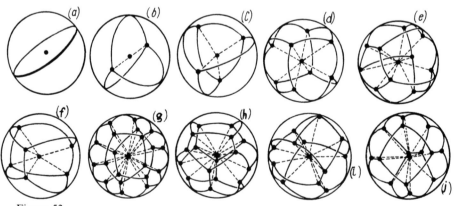

Figure 53

However, we already know very well that singular points of the film cannot be ignored and that the minimality condition places rigid restrictions on the structure of soap film singularities; especially, if stable films are of interest. The films minimal in the sense that any small perturbation with small support does not increase film area are most interesting. From the point of this definition, the majority of the cones in Fig. 53 are not minimal. First, it is clear that the first three surfaces spanning the plane circle, three arcs meeting at equal angles at two vertices, and the regular tetrahedron are stable minimal films. Second, it turns out that all the other cones are not minimal, for each of which there is a contracting deformation, or perturbation, which is area-decreasing. Meanwhile, the cone vertex decomposes, or "blows up", into topologically more complex formation, but possessing singularities fulfilling the Plateau principles. Those real soap films obtained in an attempt to realize minimal surfaces with the boundary represented in Fig. 53 are in Figs. 54, 43 [11]. These surfaces possess a sufficiently complicated structure; however, all their singular points either are placed on the triple singular edges or are four-fold singularities where four film edges meet at equal angles. Being patient and accurate enough, the interested reader can obtain real soap films of the described topological type, resorting to the wire frames in the diagram. The films are stable, and any small perturbation does not decrease their area. The film in Fig. 43 can be regarded as small no more, since it is seen not to model the structure of an infinitely small neighborhood of a singular point on the soap film. In contrast, the first three films (a), (b) and (c) admit a similitude

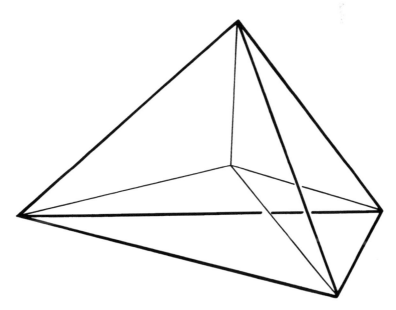

Figure 54

(contraction with the origin of coordinates as the center) sending them into themselves and decreasing their size. These films are the models of film structure in an arbitrarily small neighborhood of a singularity. Here, we have touched upon the interesting problem of minimal cones in R^n, vertex at the origin, and the "base", or boundary, given by the minimal submanifold A^{n-2} in the sphere S^{n-1}.

6. Multidimensional Minimal Cones

Consider in R^n the sphere S^{n-1} of radius R, center at the origin of coordinates. Let A^{n-2} be a $(n-2)$-dimensional compact subset of the sphere, which is a smooth $(n-2)$-dimensional submanifold of the sphere almost everywhere (except a point set of measure zero). Consider the cone CA, vertex at the point 0, and basis A. The cone is therefore formed by the radii joining the centre to the points of the set A. We fix the boundary A of CA, and assume that the cone is a minimal surface in the sense that any of its perturbations preserving the boundary increases (does not decrease) the volume, or the area in the two-dimensional case. By a cone perturbation, we mean those in Secs. 1 and 2. CA is clearly not minimal for any choice of A.

LEMMA 1. *If a cone CA is a minimal surface in R^n, then its boundary A is a minimal surface in the sphere S^{n-1} at all its regular points.*

In particular, except a point set of measure zero, the boundary A is a locally minimal submanifold of the sphere, i.e., its mean curvature is zero. It follows from the elementary properties of an integral that the $(n-1)$-dimensional volume of CA is related by the $(n-2)$-dimensional volume of A by $\text{vol}_{n-1} CA = \dfrac{R}{n-1} \text{vol}_{n-2} A$, where R is the sphere radius. Hence, any small perturbation of volume-decreasing perturbation of the boundary induces (along the radii) a similar volume-decreasing perturbation of the whole cone. Thus, if we assume that a contracting deformation for the boundary exists, then the contracting deformation also exists for the cone, which is contrary to the assumed minimality, and thus proves the lemma.

Hence, the statement already familiar to us that the intersection of a two-dimensional soap film with a sphere of arbitrarily small radius, center at a singular point of a film can be regarded as consisting of equator segments meeting at triple points at equal angles.

For simplicity, we restrict ourselves to the case where the cone boundary is a smooth and closed submanifold of the sphere. The question arises: Do there exist in R^n cones, vertices at the origin of coordinates, which are minimal in the sense that any of their small perturbations is volume-increasing? The simplest of them is the standard plane $(n-1)$-dimensional Euclidean disk passing through the origin and intersecting the sphere in the equator. We therefore sharpen the question, and ask whether or not there are non-trivial minimal cones different from the standard disk. The answer is so surprising that we are going to dwell on the problem longer. It is obvious that there are no minimal cones (save a plane disk) with a smooth one-dimensional submanifold as the boundary, which follows from the fact that a boundary should necessarily be a path of locally minimal length and that its all geodesics are exhausted by the sphere equators. Accordingly, the cone is a plane disk. As the dimension increases, locally minimal submanifolds A^{n-2} not coinciding with equatorial hyperspheres arise in S^{n-1}; e.g., a locally minimal two-dimensional torus is contained in S^3 along with the usual equators. The torus is embedded as follows: Let $S^3 \subset R^4 = C^2(z, w)$; then the torus T^2 can be specified as the intersection of the sphere $S^3 = \{|z|^2 + |w|^2 = 1\}$ with the surface $M^3 = \{|z| = |w|\}$ where the sphere is the union of two solid tori $|z| \leqslant |w|$ and $|z| \geqslant |w|$, and the torus is the common boundary. If φ and ψ are angular coordinates, then the torus pints are given by $(z, w) = (e^{i\varphi}, e^{i\varphi})/\sqrt{2}$. The induced metris is flat Euclidean, since $dzd\bar{z} + dwd\bar{w} = (d\varphi^2 + d\psi^2)/2$. It is easily verified that the torus is a locally minimal surface in the sphere. We shall return to this problem below. However, we can also see that the minimal surface inside the sphere with a torus as its boundary is not a cone. We can see that there are no non-trivial cones, centers at the origin of coordinats, either in R^4 or in R^3. As the dimension increases further, the picture alters. If the dimension is large, then it turns out that minimal non-trivial cones exist (E. Bombieri, De Georgi, E. Giusti) [22]. To get an intuitive idea of the reason why this curious effect occurs, we consider for definiteness the boundary A^{n-2} of fixed topological type. The second simplest manifolds after spheres, admitting locally minimal embeddings in the spheres, are the sphere products $S^p \times S^q$, where $p + q = n - 2$. Let $n = 2$. Consider the circle S^1, and $A = S^0 \times S^0 \subset S^1$ (Fig. 55), in which case the cone with boundary A is the union of two diameters. This one-dimensional "surface" is not minimal, while the four-fold point decomposes into the union of two triple ones. The real minimal path with boundary A is the union of two parallel line segments, or a "one-dimensional cylinder" (see the diagram). Now, let $n = 3$. We take $S^1 \times S^0$ as A in the sphere S^2. The two-dimensional cone with boundary A is evidently not minimal either, a contracting deformation occurring in the neighborhood of the vertex with a decrease in area. After this variation, the singular point, i.e., cone vertex, decomposes and turns into a circle which is the gorge of a catenoid, or a real minimal film with boundary $S^1 \times S^0$. Subjecting the two-dimensional minimal film, i.e., catenoid, to contrast with the one-dimensional, or two parallel, we notice that the two-dimensional film sags and bends in the direction of the origin unlike the one-dimensional. It so happens that the minimal film sag directed towards the origin of coordinates increases as dimension does. Consider S^3 and the above torus T^2 in it, as the boundary. The three dimensional minimal film can be computed and seen to possess still more narrower gorge than the two-

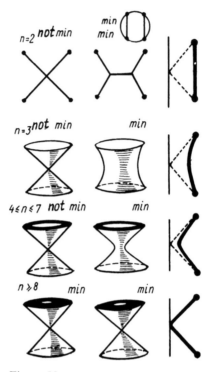

Figure 55

dimensional catenoid (see the diagram). In Fig. 55, cones with boundary $S^p \times S^q$ are represented schematically in the first column, real minimal surfaces with the same boundary in the second, and sections by planes (generators) of these minimal films by two dimensional half-planes passing through the surface axis of symmetry in the third. The section is of the form of the path sagging more and more as we approach the point 0 with an increase of dimension. Consequently, a moment comes when the minimal film sags so much in this monotonic process that its gorge collapses, contracts to a point and the film transforms into a cone, which occurs, beginning with dimension eight. For lesser dimensions inclusive, there are no minimal cones different from the plane disk (F. Almgren, J. Simons).

THEOREM 1 [23]. *Let A^{n-2} be a closed, smooth and locally minimal submanifold in the sphere S^{n-1} standardly embedded in R^n,. Suppose that A is not a totally geodesic sphere, or an equator. Then for $n \leqslant 7$, the cone CA (i.e., $(n-1)$-dimensional surface formed by all radii joining the origin of coordinates to the point of the manifold A) is not minimal, i.e., there exists a contracting deformation of the cone, which decreases its volume.*

The statement is generic in the sense that it holds also for simply connected and complete Riemannian manifolds M^n of constant sectional curvature (sphere and Lobachevski space), and not only for Euclidean space, viz., let S_R^{n-1} is a geodesic sphere in M^n, of radius R, formed by the end-points of geodesic radii of length R, emanating from a fixed point, i.e., the center of the sphere. Let CA be the cone over A, formed by the geodesic joining the center to the points of a smooth, closed and locally minimal submanifold A^{n-2} embedded in the sphere S_R^{n-1}. Then, if A is not a totally geodesic sphere, the cone CA is not minimal for $n \leqslant 7$. Moreover, this statement holds true for any Riemannian spaces derived from R^n by multiplying the Euclidean metric by a positive function $f(R)$ dependent only on R (proved by the Moscow University student Khorkova).

We now come back to the case of R^n. Starting with $n = 8$, there are now in R^n cones which are also globally minimal, and not only locally. We illustrate by example. Consider the product of spheres $S^3 \times S^3$ of the same radius as the boundary, naturally embedded in S^7 as a locally minimal submanifold [23, 24]. To indicate the embedding, we consider in $\mathbf{R}^{2p}$ the sphere S^{2p-1} of unit radius $x_1^2 + \ldots + x_{2p}^2 = 1$ and intersect it by the cone $x_1^2 + \ldots + x_p^2 = x_{p+1}^2 + \ldots + x_{2p}^2$. The intersection is the diffeomorphic to $S^{p-1} \times S^{p-1}$. We see by computation that the submanifold $S^{p-1} \times S^{p-1}$ is locally minimal in the sphere. We shall see that it is true in the sequel on the basis of other argument.

7. Minimal Surfaces Invariant under the Action of Lie Groups

Consider a smooth Riemannian manifold M with its isometry group acting on it smoothly. Let I_0 be the connected component of the identity element of this Lie group, and G a certain connected and compact subgroup in I_0, and $X \subset M$ a surface invariant under the action of the group G. Such surfaces are said to be G-invariant. They fiber into the orbits of action of G. Consider the multidimensional problem of Plateau. Let a "boundary frame", a closed submanifold A^{n-1} invariant under the action of G, be fixed in M. It is natural to expect that if the boundary has a symmetry group (or G-invariant), then the minimal surface spanning the boundary should also possess the same group. The equivariant Plateau problem consists in ascertaining the cases where the existence of a G-invariant minimal surface X^{n-1} can be guaranteed, spanning the G-invariant boundary (in the above senses).

By means of differential geometry, it can be proved that a G-invariant submanifold X of a manifold M is locally minimal with respect to all sufficiently small variations if and only if it is locally minimal only with respect to all sufficiently small *equivariant variations* (i.e., those invariant under the action of the same group) (see, e.g., [25]). Thus, to check that a *G-invariant surface* is locally minimal (in the sense that mean curvature is zero), it suffices to see that the surface is minimal with respect to a narrower variation class, viz., G-invariant perturbations, which enables us, as shown in the sequel, to reduce the problem of finding G-invariant minimal surfaces in M to the problem of investigation of the minimal surfaces in the orbit space M/G.

Consider the decomposition of the manifold M into the union of orbits of $G(x)$ of action of G, where $x \in M$. Generally speaking, orbits of different dimensions "grow" at different points; however, a certain subclass of orbits said to be generic, or principal, can be naturally distinguished in the set of all orbits. To describe it, recall that each orbit $G(x)$ admits a representation in the form of a homogeneous space $G/H(x)$, where $H(x)$ is the stability subgroup of the point x, i.e., the collection of all transformations from G, leaving the point fixed. The set of all cosets in G relative to $H(x)$ is just what is identified with the space $G/H(x)$. We show that such a maximal, open and everywhere dense subset $\tilde{M}$ can be distinguished in M, which is in fact an open and smooth submanifold of M that, for any two points x, y from this subset, the corresponding stability subgroups $H(x)$ and $H(y)$ are conjugate in G, i.e., there is an element $g \in G$, so that $H(y) = gH(x)g^{-1}$. The orbits $G(x)$ corresponding to the points x from this everywhere dense $\tilde{M}$ are accordingly termed principal, or generic; in particular, all of them are pairwise diffeomorphic, and of the same dimension, which is maximum in the class of all orbits. All the others are said to be singular. The set of singular orbits is of measure zero in the space of all orbits, and the corresponding points from M fill up a closed subset whose n-dimensional volume (measure) is zero, where $n = \dim M$; G is assumed to be compact. There can exist singular orbits of maximum dimension, which are not principal; however, since they fill up the subset of measure zero, they are not of interest in the sequel.

Let X^q be an arbitrary G-invariant surface in M. Put $k = q - s$, where s is the dimension of a principal orbit, i.e., k is the codimension of generic orbits in the surface X. Since each principal orbit $a = G(x)$ is a smooth s-dimensional submanifold of M, we can compute its s-dimensional volume which we denote by $v(a)$, where a is the orbit $G(x)$) passing through a point x. Consider the orbit space M/G. Generally speaking, it is not a manifold, and can contain singularities; however, it contains everywhere dense subset $\tilde{M}/G$ which is a manifold. Let $\pi : M \to M/G$ be the projection of M onto M/G, associating each point x with the orbit $G(x) = a$ passing through it. The restriction of π to an open submanifold $\tilde{M}$ of M defines a smooth fibration with fiber $\pi^{-1}(a) = G(x)$ over the base space $\tilde{M}/G$ (see Sec. 3, Ch. 1). Though all fibres $\pi^{-1}(a)$ over the base space M/G are diffeomorphic, generally speaking, they are of different s-dimensional volumes if the fibres are regarded as submanifolds $G(x)$ of M. Therefore, assigning its volume to each orbit, we obtain a smooth function $v(a)$ defined on $\tilde{M}/G$ (Fig. 56), and important in the sequel [25, 26] (W. Hsiang and H. Lawson).

We now define the Riemannian metric on $\tilde{M}/G$, which will be useful in the sequel. We resort to the Riemannian metric ds specified on M, and the orbit volume function $v(a)$. Let $\langle b, c \rangle$ be the scalar product of tangent vectors b and c, determined by the metric on M. Fix the set of $(n-s)$-dimensional planes in the tangent planes to M, and orthogonal to the orbit $G(x)$, i.e., fibers of the fiber space $\pi: \tilde{M} \to M/G$. Let b' and c' be two vectors tangent to $\tilde{M}/G$ at the point a. Their inverse images, or two vectors b and c in the plane orthogonal to $G(x)$, and projecting into b' and c' under the mapping $d\pi_x$ onto the b' and c' are obviously uniquely determined at each point $x = \pi^{-1}(a)$ (Fig. 56). We now specify non-singular symmetric scalar product $\langle , \rangle$ of b' and c' by putting $\langle b', c' \rangle' = \langle b, c \rangle$. We have thus defined the Riemannian metric on $\tilde{M}/G$, which is the projection of the original metric ds. Denote the obtained metric by $d\tilde{s}$. Finally, we define the metric dl on $\tilde{M}/G$ by putting $dl = v(a)^{1/k} d\tilde{s}$ at each point a from $\tilde{M}/G$. Since $v(a)$ is a positive smooth function on $\tilde{M}/G$, dl is a non-singular positive definite metric on the quotient manifold $\tilde{M}/G$. In fact, we can extend $v(a)$ to a continuous function on the whole of $\tilde{M}/G$, defining $v(a)$ at all points a which are singular orbits; however, we are not going to use this circumstance.

The choice of dl is due to one simple but important statement. Let X be a compact and G-invariant submanifold of M. Consider the quotion space X/G embedded in M/G. We assume that X is in general position with respect to the set of singular orbits of M, and consider in X the subset $\tilde{X} = X \cap \tilde{M}$, of those points with principal orbits. Since X is a G-invariant surface, it is representable as the union of orbits. We assume that almost all of them are principal, i.e., $\tilde{X}$ is an open and everywhere dense submanifold of X; in particular $\mathrm{vol}_q X = \mathrm{vol}_q \tilde{X}$. Recall that k is the codimension of a principal orbit in X; therefore k is the dimension of the quotient space X/G.

LEMMA 2. *If dl is the metric introduced above on $\tilde{M}/G$, then the q-dimensional volume of the surface X^q in M, G-invariant with respect to the metric ds is equal to the $\tilde{K}$-dimensional volume of the quotient surface $\tilde{X}/G$ in the quotient space $\tilde{M}/G$ with respect to the dl, where $k = \dim X - \dim G(x)$, $G(x)$ is a principal orbit.*

Proof follows from simple geometric arguments. In fact, the k-dimensional volume (with respect to dl) of an elementary infinitely arbitrarily small cube Δ^k with edge dl, placed in $\tilde{X}/G$, is $(dl)^k$, where dl also schematically represents the edge length; hence, $\mathrm{vol}_k \Delta^k = v(a)(d\tilde{s})^k$. It also follows from Fig. 56(b) that this quantity equals the product of the volume of the base of the parallelepiped σ^q and the volume of its "generator". Meanwhile, the parallelepiped base is the element of area of dimension k, whose volume is obviously $(d\tilde{s})^k$. The generator of σ^q is the orbit of group action, while the volume of $G(x)$ is equal to $v(a)$, where $a = \pi(x)$.

Hence, the following.

PROPOSITION 1. *Let X^q be a G-invariant minimal submanifold of M. It is locally minimal (i.e., its mean curvature is zero) in M with respect to the metric ds if and* *only if the quotient manifold $\tilde{X}/G$ is locally minimal in the quotient manifold $\tilde{M}/G$ with respect to the metric $dl = v^{1/k} \cdot d\tilde{s}$, where $k = \dim X - \dim G(x)$, i.e., the value k is the codimension of a generic orbit in the surface X.*

As early as $k = 0$, we derive some interesting corollaries. If X^q is for simplicity regarded as connected, then equality of k to zero means that X is a certain orbit of action of the group G on the ambient manifold M. Turning to the proof of Lemma 2 again, we see that the parallelepiped degenerates here into its generator, i.e., orbit $G(x)$ coinciding with X. Therefore, the local minimality of X in M with respect to ds is equivalent to the condition that the volume function v attains an extremum at the point $a = \pi(X) = \pi G(x)$. Thus, all locally minimal principal orbits $G(x)$ in M are to be found by first considering the space of principal orbits $\tilde{M}/G$, computing $v(a)$ on it and finding all its critical points, i.e., those at which grad $v(a) = 0$. They are just the orbits which are locally minimal submanifolds of M, the simplest example being the action of the one-dimensional group $G = SO(2)$ as the rotation group of the standard sphere S^2 about the vertical axis. The orbit space S^2/G is here one-dimensional, and homeomorphic to a line segment, whereas the principal orbit space is homeomorphic to an open interval. The volume function is defined on an open interval, is the square root of the quadratic function and has only one extremum, viz, maximum attained at the center of the interval. The corresponding orbit is a circle of greatest radius, or equator. Therefore, an equator is a geodesic in the sphere.

For $k = 1$, a G-invariant submanifold X is locally minimal in M with respect to ds if and only if the path $\tilde{X}/G$ is a geodesic in $\tilde{M}/G$ with respect to the metric $dl = v(a)d\tilde{s}$. The case is of special interest, and analyzed in the sequel. Now, we consider the equivariant Plateau problem.

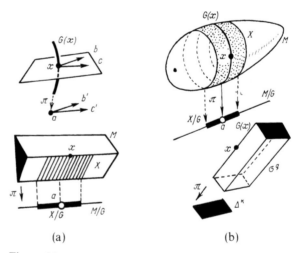

(a)   (b)

Figure 56

THEOREM 2 (H. Lawson [26]). *Let G be a closed subgroup of the group of proper orthogonal rotations R^n, i.e., $G \subset SO(n)$, A^{n-2} a smooth manifold of the standard sphere S^{n-1}, invariant with respect to the action of the group G. Then there exists a minimal surface (in the generalized sense) X_0^{n-1} with boundary A^{n-2}, invariant under the action of G. If this G-invariant surface X_0^{n-1} is unique, then it is also mechanically globally minimal in the class of all surfaces with given boundary, but now not necessarily G-invariant.*

Thus, if the boundary admits a non-trivial symmetry group G, then, to find an absolutely minimal surface with this boundary, it sometimes suffices to find a minimal surface only in the class of G-invariant films. The concept of surface and its boundary in the theorem are sharpened, e.g., in [26]. To drop the requirement that the transformations of G should preserve the orientation of R^n (see Theorem 2) is impossible as simple examples demonstrate. Theorem 2 was expanded to the case of Riemannian manifolds (replacing R^n by M^n) by J. Brothers in [31].

8. Fermat Principle, Minimal Cones and Light Rays

Considering the cone problem again, we assume that the connected boundary $A^{n-2} \subset S^{n-1}$ is a principal (generic) orbit of the action of a compact Lie group G on R^n. Under the projection π onto R/G, A^{n-2} is carried into a point $a = \pi A$. The cone CA is also a G-invariant surface, and projected by the mapping $\pi : R^n \to R^n/G$ into the one-dimensional path joining a and 0 together, where 0 is the image of the singular orbit, or the origin of coordinates.

When is CA globally minimal, i.e., when does it realize the volume functional absolute minimum in the class of all surfaces with the same boundary? We know that if the dimension n is small, then the cone is not minimal. Taking into account Sec. 7, we should find in $\tilde{R}^n/G$ the shortest geodesic emanating from a, and having its end-point on the boundary of $\tilde{R}^n/G$. If the geodesic coincides with the path which is the cone projection, then the cone is globally minimal. A priori, we study the structure of R^n/G, and consider the example of $A^{n-2} = S^{p-1} \times S^{p-1}$, where $n = 2p$; here

$$A^{n-2} = \left(\sum_{i=1}^{p} x_i^2 + y_i^2 = 1 \right) \cap \left(\sum_{i=1}^{p} x_i^2 = \sum_{i=1}^{p} y_i^2 \right),$$

As G we take the boundary of $SO(p) \times SO(p)$, which is a subgroup of $SO(2p)$, the first component of $SO(p)$ in fact acting on $R^p(x_1, \ldots, x_p)$, while the second on $R^p(y_1, \ldots, y_p)$. It is obvious that the quotient space R^{2p}/G is two-dimensional, because the group $SO(2p)$ acts on S^{p-1} transitively, i.e., any point of the sphere can be mapped into any by group transformations. Since orthogonal transformations preserve the moduli of vectors, we can take the pair of numbers x, y, i.e., the lengths of the vectors $(x_1, \ldots, x_p)$ and $(y_1, \ldots, y_p)$, as coordinates on $\mathbf{R}^{2p}/G$. Since $\mathbf{R}^p(x_1,\ldots,x_p)/SO(p)$ *is isomorphic to the half-line* $x \geqslant 0$ *and* $\mathbf{R}^p(y_1,\ldots,y_p)/SO(p)$ to the half-line $y \geqslant 0$, $\mathbf{R}^{2p}/G$ is isomorphic to the set of points of the two-dimensional plane determined by the inequalities $x \geqslant 0$, $y \geqslant 0$, i.e., the first quadrant. We now find the metric $d\tilde{s}$ induced on $\mathbf{R}^{2p}/G$ by the standard Euclidean metric ds on $\mathbf{R}^{2p}$. Consider the element of area orthogonal to the generic orbit $S^{p-1} \times S^{p-1}$ at its arbitrary point. The scalar product arising on this element and enabling us to measure the distance between infinitely arbitrarily close orbits is clearly generated by the two-dimensional Euclidean metric, for the group G acts orthogonally. To find the volume function $v(a)$ on the orbit space, we have to calculate the $(2p-2)$-dimensional volume of the direct product of the two spheres $S^{p-1}(x) \times S^{p-1}(y)$ of radii x and y, respectively. Since the volume of a $(p-1)$ sphere of radius x is αx^{p-1}, where α is a certain constant $v(a) = \alpha^2(xy)^{p-1}$. Thus, we can now calculate the metric dl on $\mathbf{R}^{2p}/G$. Up to a constant non-zero multiplier, we have $dl^2 = (xy)^{2p-2}(dx^2 + dy^2)$. To find the locally minimal submanifold A^{2p-2} of S^{2p-1}, which is the generic orbit of the above action of G, we should find the volume function $v(a)$ extrema on $S^{2p-1}/G \subset \mathbf{R}^2/G$ whose explicit form was found above. It is obvious that the sphere S^{2p-1} is projected by the mapping π onto an arc of the circle $x^2 + y^2 = 1$, placed in the first quadrant. Thus, it remains to find the volume function extrema, if the function is restricted to this circle. We have $v = (xy)^{p-1} = x^{p-1} \cdot (1-x^2)^{(p-1)/2}$. The function clearly possesses three extrema $x = 0$, 1, $1/\sqrt{2}$ with only one of interest, and corresponding to the generic orbit, viz., $x = 1/\sqrt{2}$. Thus, the locally minimal orbit is on the first quadrant bisector.

To summarize, the quotient space $\mathbf{R}^{2p}/G$ is the closed first quadrant of the two-dimensional plane, the metric dl^2 is of the form $(xy)^{2p-2}(dx^2 + dy^2)$, the locally minimal submanifold $A = S^{p-1} \times S^{p-1}$ embedded in the sphere S^{2p-1} is represented as a point a on the first quadrant bisector, and the cone CA is given by the bisector segment joining a to the origin of coordinates. Our goal is now to find the geodesic of least length from a point onto the boundary of the domain $\tilde{R}^n/G$.

The following mechanical analogy is of use in the sequel. The function $v(a) = n(x,y)$ is smooth on $\tilde{R}^n/G$, and vanishes on the boundary. Consider a two-dimensional solid and transparent medium filling up the first quadrant on the Euclidean plane, with refraction constant $n(x,y) = c/w(x,y)$, where c is light velocity in vacuum and $w(x,y)$ is the velocity of light in the medium at a point with coordinates (x,y). The transparent medium is assumed to be isotropic at each point. However, the medium is not homogeneous if the refraction constant is not invariable. The Fermat principle states that a light ray spreading in the medium from a point P to a point B travels in a path minimizing for fixed energy the time to pass from P and B. Furthermore, the paths in which light rays travel in the medium with refraction index $n(x,y)$ are at the same time geodesics with respect to the conformal metric

$n(x, y)(dx^2 + dy^2)$. The converse is also true; therefore, to describe the behavior of the beam of light rays from a point-source, it suffices to describe the structure of the pencil of geodesics emanating from this point in all directions. Consider the particular case where the medium fills up the upper half-plane (over the Ox-axis) and the refraction index in a smooth function vanishing on the domain boundary, not depending on the coordinate x and monotonically increasing as y increases, i.e., $n(x, y) = n(y)$, $n(o) = 0$, $n(y_1) > n(y_2)$ when $y_1 > y_2$. Let a narrow beam of light rays pass from a point inside the medium towards the boundary, or the horizontal axis. The qualitative picture is in Fig. 57. The vertical ray reaches the boundary orthogonally to the axis, while close rays deviate to the right and left, which we shall make use of soon. We now turn to the case of interest, where $n(x, y) = (xy)^{p-1}$. It follows from the function symmetry with respect to coordinate change that the bisector is a geodesic; therefore, the cone CA is locally minimal at all its non-singular point, i.e., of zero mean curvature for all $p \geqslant 1$. Note that this is not contrary to the information that there are no minimal cones other than plane disks in spaces R^n of small dimensions. As a matter of fact, that a cone is locally minimal in the sense of differential geometry (i.e., minimal with respect to small variations with small support, concentrated near a regular point) is not an obstacle to the existence of contracting deformations with sufficiently large support or the existence of small contracting deformations in the vicinity of singulary points. This is just what occurs in small dimensions; e.g., $A^0 = S^0 \times S^0$ when $p = 1$, the situation is trivial and familiar, where $S^0 \times S^0 \subset S^1 \subset \mathbf{R}^2$ and the cone CA consists of two diameters meeting at the origin at right angles. Though each is a geodesic, there is a contracting deformation at the one-dimensional cone vertex, leading to the decomposition of the four-fold point into the unin of two triple ones. The truly minimal one-dimensional surface consists here of vertical line segments (Fig. 55). From the standpoint of the last model on the

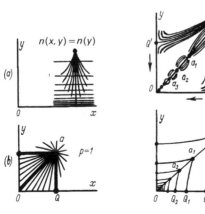

Figure 57

two-dimensional plane, the metric dl for $p = 1$ is of the form $dx^2 + dy^2$, and the picture for the rays emanating from a point a is as in Fig. 57(b) is the same as the pencil of straight lines from the point a. It is clear that the bisector is not the shortest geodesic, while the line segment aQ orthogonally meeting the Ox-axis is. It is just a generator of the one-dimensional cylinder (Fig. 55). Finally, we just obtain the interpretation of the paths given in the last column in the diagram. The upper path of the curves turns out to represent the least geodesic joining the point source a to the boundary of the domain $\hat{\mathbf{R}}^n/G$ coinciding with part of the plane.

Let $p > 1$. For simplicity, we regard p as a continuous parameter. In spite of p originally having represented integral dimension, from the point of view of analysis of geodesics relative to the metric dl, it is not necessary to restrict ourselves only to the whole parameter values. Moreover, the sharp qualitative change in which we are interested, or changing situation with the geodesics just occurs for a cricial value p_0 which is not an integer.

It turns out that, as p increases, the vertical line segment aQ representing the least geodesic when $p = 1$ starts bending in the direction of the origin, and, which is especially important, remains the least geodesic joining the point a to the domain boundary. The point Q begins shifting monotonically in the direction of the point 0. Due to the symmetry of dl with respect to permutations of coordinates x, y, the same also occurs over the bisector. For $p > 1$, the situation gets more complicated. The qualitative picture of the light beam from a is in Fig. 57(c) when $p > 1$. It is interesting that, in the segment of the bisector from a until the origin, infinitely many brightly illuminated points $a_1, a_2, a_3, \ldots$ arise (for the values of p close to unity) at which close geodesics are collected and organized into narrow pencil along the bisector towards the origin. The first of these is a with all geodesics arbitrarily close to the bisector. Such points are said to be conjugated. Straightforward calculation shows that there are infinitely many such points. The qualitative argument predicting just such a picture follows from the above remark regarding the nature of refraction of light rays traveling towards the medium boundary in the direction of decreasing refraction index, which clearly accounts for the behavior of geodesics close to the path aQ (Fig. 57(c)). Analogous reasoning involving light refraction angles can be made use of to predict the availability of conjugate points along the bisector to sufficiently small values of $p > 1$, though an exact critical value p_0 for dimension p after which the situation changes cannot be indicated; e.g., similar calculations are given in [27], and not only for the case under consideration; for other examples of minimal cones, too. It is clear that if a point a_1 is taken as the source and not a, then the representation of light rays from it is identical with the Fig. 57(c); in particular, a geodesic a_iQ_i different from the bisector at right angles to the boundary emanates from each point a_i, and, which is important, is the least geodesic joining a to the boundary (Fig. 57(d)). Since, for $k = 1$, the length of the path

$\tilde{X}/G$ is the volume of the corresponding surface $X \subset M$, the cone is not globally minimal for small $p(p > 1)$. That the contracting deformation exists can be seen in Fig. 57. Consider an arbitrary point a_i conjugate to a along the geodesic bisector. Since geodesics meet at this point, arbitrarily close to the bisector and emanated from the point source a, by Morse theory [1,2], there exist a deformation decreasing the length of a geodesic in the neighborhood of a_i. A conracting deformation arises also for the corresponding sources, which proved that the cone is not minimal. With small support on $\mathbf{R}^n/G$, this deformation possesses this property no more in transferring to the original surface in $\mathbf{R}^n$, since the constructed equivariant deformation should occur along the whole orbit associated with a_i. At the same time, we can construct a contracting deformation with arbitrarily small support and small amplitude; however, this deformation is concentrated in the vicinity of the cone vertex (Fig. 57(d)). As the "perturbed" cone, we take the surface invariant under the action of the group G and represented on the quotient space $\mathbf{R}^n/G$ as follows. The path simply coincides with the bisector segment from a to a_i where the subscript i has been chosen to be sufficiently large; farther off from a until the domain boundary, the path coincides with the shortest geodesic $a_i Q_i$ emanating onto the horizontal axis. The complete path made up of two geodesic segments obviously is of strictly lesser length (for sufficiently small $p > 1$) strictly less than the bisector segment from a to 0, representing the cone CA. In this perturbation, the cone vertex decomposes and turns into the sphere S^{p-1} of small radius, which is the orbit of the group G. It "grows" from the point Q_i on the Ox-axis. That the "vanishing" cycle S^{p-1} appears is already familiar to us from the above examples (Fig. 55). The constructed perturbation is not a continuous cone deformation; however, both (original and perturbed) surfaces possess the same boundary; e.g., in the homology sense, though their topology type is different.

In the previous section, we encountered the situation where there is an unstable surface in the functional space of all surfaces with given boundary, i.e., the surface is a critical point for the volume functional, and of positive index. This means that, in the space of all surfaces, there are directions so that perturbations along them (arbitrarily small in amplitude) decrease volume of the surface. From the point of view of geometry, the graph of the volume functional in a small neighborhood of this extremal film is of saddle form. The number of such independent directions is what is called an index. For cones of small dimensions, they turn out to be also unstable extremal surfaces with infinite index at all their non-singular points, besides being locally minimal. In other words, there are infinitely many independent perturbation directions along which the cone volume starts decreasing. These deformations have actually been constructed by us. Let a_i be an arbitrary point conjugated to the point a along the bisector (Fig. 58); therefore, there exists an arbitrarily small rotation of the geodesic $s_1{}^1$, joining a and a_i together into a close position s_2, a

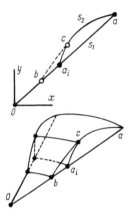

Figure 58

a_i remaining fixed and rotating geodesic length remaining fixed (up to small higher orders). Let b and c be two points close to a_i, and paced as in Fig. 58. Consider the arbitrarily small triangle bca_i. Since the metric is positive definite, owing to the triangle inequality, the length bc is less than ba_i and a_ic together, i.e., the length of the broken line $acb0$ is strictly less than the length of the broken line aca_ib0. However, the length of the latter equals that of the line segment $a0$, since the lengths of aa_i and aca_i are the same. We have thus constructed an extremal length decreasing perturbation. The corresponding cone perturbation generated by the bisector rotation is represented in Fig. 58. It is clear that there are infinitely many such independent perturbations, since there are infinitely many points a_i conjugate to a.

We now make the next step, and begin increasing the dimension p. The minimal geodesic aQ meanwhile starts moving towards the bisector. The quadrilateral $aQ0Q'$ in Fig. 57(c) gets contracted onto the bisector segment $a0$, and, finally, as p reaches a critical value, collapses onto the segment turning into the geodesic of least length joining a to the boundary. Though precise computations are complicated (and omitted for this reason), we can assert that the described situation occurs for $3.6 < p_0 < 4$. For all $p \geqslant 4$, the qualitative picture of the beam of light from the point a is essentially different from the one studied (Fig. 59). As p increases further, the picture remains unaltered.

Thus, for $p \geqslant 4$, non-trivial globally minimal cones appear, the first example being the cone over $S^3 \times S^3 \subset S^7 \subset R^8$ for $p = 4$. Recall that, since for $p \geqslant 4$, there exists a unique minimal geodesic emanating onto the domain boundary from A and coinciding with the bisector, its inverse image under the projection π is a G-invariant surface. By its uniqueness and Theorem 2, it follows that it is globally minimal also in the class of all variations, now not necessarily equivariant.

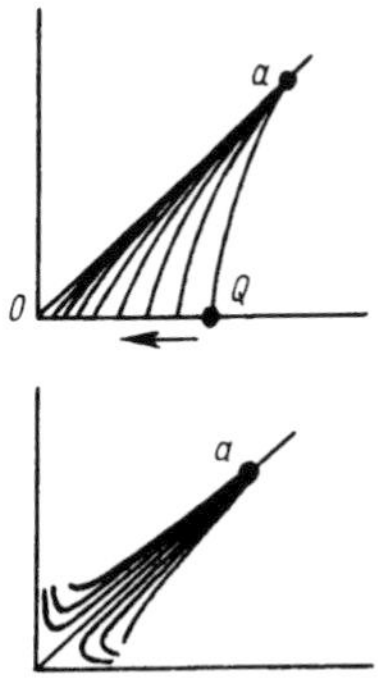

Figure 59

The direct product of spheres is not a unique manifold over which there are globally minimal cones [25–27].

Let G be a compact Lie group with adjoint action on its Lie algebra L. The sphere of unit radius $S \subset L$ is invariant relative to this action, and fibered into orbits. Let us consider generic orbits $0(x) \subset S$, and cones over these orbits, vertices at the origin. In [27], the following was posed.

PROBLEM 3. *Check these cones for minimality, viz., find all locally minimal cones over generic orbits, and clarify their stability and global minimality.*

Considerable progress in solving this problem was made by I. S. Balinskaya.

Since the cones under discussion are G-invariant, we make use of general methods for the Plateau equivariant problem.

Let $H \subset L$ be a Cartan subalgebra, and W a Weyl group. The orbit space of adjoint action of the group G coincides with H/W, and is identified with a fixed Weyl chamber $V \subset H$.

PROPOSITION . Consider the metric $dl^2 = V^2(x)d\tilde{S}^2$ on V, where $v(x)$ is the volume of the orbit $0(x)$ and $d\tilde{S}^2$ is the metric given by the Killing form. Let $x_0 \in V$, $|x_0| = 1$. (i) The cone over the orbit $0(x)$ is locally minimal $\Leftrightarrow$ the segment joining x_0 to the origin is a geodesic in the metric dl^2 $\Leftrightarrow$ the function v restricted to the sphere has an extremum at the point x_0. (ii) The cone over the orbit $0(x)$ is stable $\Leftrightarrow$ the geodesic $0x_0$ is stable. (iii) The cone over the orbit $0(x)$ is globally minimal $\Leftrightarrow$ geodesic $0x_0$ is the shortest. Thus, at first, we should find all extrema of the function $v(x)$ on the sphere of unit radius in the Cartan subalgebra.

PROPOSITION ([83]–[85]). *In each Weyl chamber, there is precisely one minimum of the volume function $v(x)$ restricted to the sphere.*

Explicit form of the volume function is used in the proof, viz.,

$$v(x) = \prod_{\alpha \in \Sigma^+} (h_\alpha, x)^2.$$

Thus, there is precisely one locally minimal cone over a generic orbit. Checking this cone for stability is reduced to estimating the norm of the Hessian of the volume function on the sphere at the extremum point x_0. For classical series, this work has been done by I. S. Balinskaya, who obtained the following result.

THEOREM. *In the case of the Lie groups $SU(k), k \geqslant 5, SO(k), k \geqslant 8, Sp(2k), k \geqslant 4$, the unique locally minimal cone is stable. For the groups $SU(k)$, $k < 5$, $SO(k)$, $k < 8$, $Sp(2k)$, $k < 4$, this cone is unstable.*

Global minimality of cones is proved by I. S. Balinskaya by means of the calibration form $\varphi = df$. In particular, some ideas developed by A. V. Bolsinov are used. The function f of the following form may be taken

$$
f(x) = \begin{cases} \dfrac{r^{N+1}}{N+1} \; v(x_0)\cos^2\!\left(\dfrac{N+1}{2}\right)\varphi \\[2mm] \qquad \text{if} \quad \varphi < \dfrac{\pi}{N+1} \\[4mm] 0 \qquad \text{if} \quad \varphi \geqslant \dfrac{\pi}{N+1} \end{cases} ,
$$

where r is the distance from the point x to the origin (in the Killing sense), φ is the angle between x and x_0, x_0 is an extremum point, and N is the orbit dimension (Fig. 60).

We should verify two conditions: (i) $|\text{grad } f(x)| \leqslant v(x)$ (grad $f(x)$ and its norm are taken in the sense of ds^2; (ii) at the points of the form $\lambda x_0 (\lambda \in [0, +\infty))$, i.e., on the geodesic line, grad f is along the line and $|\text{grad } f(\lambda x_0)| = v(\lambda x_0)$. Condition (ii) holds mechanically.

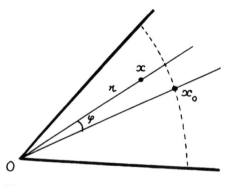

Figure 60

Let us check condition (i). A direct computation shows that

$$|\text{grad } f(x)| = \begin{cases} r^N v(x_0) \cos\left(\dfrac{N+1}{2}\right)\varphi & \text{if } \varphi < \dfrac{\pi}{N+1} \\ 0 & \text{if } \varphi \geqslant \dfrac{\pi}{N+1} \end{cases},$$

whereupon it follows that it suffices to verify the condition $v(x_0)\cos\left(\dfrac{N+1}{2}\right)\varphi \leqslant v(x)$ for $\varphi < \dfrac{\pi}{N+1}$, i.e., in a small ball only on the sphere of unit radius.

First of all, we should make sure that, for $\varphi < \dfrac{\pi}{N+1}$, the point is contained in the Weyl chamber. Then we can verify the condition $v(x_0)\cos\left(\dfrac{N+1}{2}\right)\varphi \leqslant v(x)$ on any geodesic $\gamma(t)$ on the sphere, emanating from the point $x_0 = \gamma(0)$. At the point $x_0 = \gamma(0)$, we have $v(x_0)\cos\left(\dfrac{N+1}{2}\right)\varphi \leqslant v(x_0)$, since $\varphi = 0$ and $\dfrac{d}{dt}\Big|_{t=0} v(x_0)\cos\left(\dfrac{N+1}{2}\right)\varphi(t) = \dfrac{d}{dt}\Big|_{t=0} v(\gamma(t)) = 0$. Therefore, the second derivative $\dfrac{d^2}{dt^2}$ of these two functions is estimated on $\gamma(t)$.

It turns out that, for this end, it suffices to show that, for $\varphi < \dfrac{\pi}{N+1}$, the point is distant enough from the wall of the Weyl chamber. The following theorem is proved by this method.

THEOREM (due to I. S. Balinskaya). *For sufficiently large n, the unique locally minimal cone over a generic orbit of adjoint representation of the group SU(n) is globally minimal.*

9. S. N. Bernshstein Problem

It is so clearly related to the cone problem that we are going to mention it just here. Consider in R^n the graph over a smooth function $x_n = f(x_1, \ldots, x_{n-1})$ defined on the whole hyperplane R^{n-1}. The graph is a smooth submanifold in R^n. Assume that it is locally minimal. The question arises: Is then the function f linear, and its graph a hyperplane? S. N. Bernhstein has answered positively for $n = 3$. However, the reasoning was of no avail for $n > 3$; therefore, it was required to develop new mathematical machinery. It was obtained that the final answer depended on dimension n [22, 23]. It was found that, for small n, any locally minimal graph is linear, while, for large n, there were essentially non-linear graphs locally and even globally, minimal.

The pretext for considering this problem was originally given by the physical experiments in the 19th century over soap films. It had been noticed long ago that if a closed contour had been considered in R^3, admitting the one-to-one projection onto a fixed two-dimensional plane as a convex piecewise smooth curve, then the contour could always span a stable soap film homeomorphic to the disk. If the contour was extended, leaving its projection convex, then the soap disk started deforming continuously, increasing its size. The experiments show that, as the contour increases (to the limits specified by the force of gravity), the soap disk starts flattening in its central part, gradually turning into a surface close to the plane (shown by J. Plateau). We can see that the flattening starts always, and is not related to the extent of a contour complexity. Thus, the hypothesis arises that, if the contour is at infinity, then the soap film spanning it should gradually flatten far from the contour and tend to a plane. Thus, from the experimenter's point of view in the center near to the origin, the soap film by degrees gets indistinguishable from the usual plane. It is assumed meanwhile that the observer possesses limited possibilities, and can see only a bounded volume in space. The reduction of the minimal graph problem to the one described is made on the basis of the Plateau principle, according to which part of the soap film bounded by any closed curve represented on the film also turns out to be a minimal surface with respect to this contour. The exhaustive reply is given by the following result (due to E. Bombieri, De Georgi, E. Giusti, F. Almgren, J. Simons).

THEOREM 3 [22]. *Let $x_n = f(x_1, \ldots, x_{n-1})$ be a smooth function defined on the whole hyperplane R^{n-1} in R^n, and its graph locally minimal, i.e., of zero mean curvature. The function f is then linear for $n \leqslant 8$. However, if $n \geqslant 9$, then there exist non-linear functions whose graphs are even globally minimal surfaces, and not only locally.*

The global minimality is understood here in the following sense: If we consider an arbitrary continuous perturbation with compact support, leaving the surface fixed outside a compact domain, then its volume does not decrease. A minimal cone is naturally related to each minimal graph. It turns out that it follows from local graph minimality that it is globally minimal. For simplicity, let the graph pass through the origin of coordinates. Consider the ball of radius R, with center at the point 0. Suppose that Γ_R is the part of the graph bounded by the ball. Changing the scale, i.e., effecting a similitude with coefficient $1/R$, contract the ball D_R onto D_1 of unit radius. The surface Γ_R then contracts and turns into the one denoted by M_R. The latter depends on R. As R increases, the surface somehow alters inside the ball D_1, its boundary always remaining on that of the ball. Consider the "limit" of minimal surfaces M_R as $R \to \infty$. This limit exists (reasonably) and is the cone CA over a certain $(n-2)$-dimensional subset A in the sphere of radius 1. It can be proved that A is a locally minimal submanifold, the cone being globally minimal for fixed boundary. With the familiar dimensional restrictions placed for $n \leqslant 7$, the cone should be a plane disk. It is deduced that the original graph should also be a hyperplane, which proves the theorem. When $n = 8$, certain extra argument is required [22]. The cone construction from the graph can be illustrated visually if we imagine an observer who, in contrast to the prior, tends to infinity, leaving the origin, and consider the graph from the point at infinity. All the small oscillations of the graph which are not conical then "vanish" at the origin.

That non-linear graphs exist for $n \geqslant 9$ is proved analytically though the same minimal cones are involved [22].

10. Complex Submanifolds as Minimal Surfaces

There is a wide class of minimal surfaces often encountered in geometry. These are complex submanifolds of a Kählerian manifold M. Recall that a complex manifold is said to be Kählerian if its metric $ds^2 = \sum\limits_{i,j} g_{ij} dz_i d\bar{z}_i$ determines a closed exterior differential for $\omega = \frac{i}{2} \prod\limits_{i,j} g_{ij} dz_i \wedge d\bar{z}_j$, where $z_1, \ldots, z_n$ are local coordinates and the bar denotes complex conjugation. Kählerian manifolds are also given by the complex linear space C^n and projective space CP^n.

We are already familiar with several notions of variation of a surface. In the simplest cases, this is a continuous deformation, or surface homotopy. In an algebraic approach, it is the transition from one cycle to another homologous to the former. We now indicate another natural concept of variation often arising in applications. Let $X^k \subset M^n$ be a smooth, compact, orientable and closed manifolds. We will say that its bordism deformation is given if a $(k-1)$-dimensional smooth, compact and orientable submanifold Z with boundary $\partial Z = X \cup (-Y)$ is given, where Y is a certain smooth, compact and orientable k-dimensional submanifold of M, and Y with opposite orientation. We call U the *bordism-variation* of the original X. In the case where X is non-compact, we will say that its bordism deformation is given if a submanifold Y coinciding with X outside a compact domain is given and, moreover, a $(k-1)$-dimensional submanifold Z with piecewise smooth boundary $\partial Z \subset X \cup (-Y)$ is given (Fig. 61, 62). From the homology stand-

Figure 61

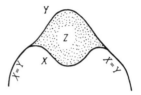

Figure 62

point, the two cycles $[X]$ and $[Y]$ realized by X and Y are clearly homologous; however, the above equivalence relation sometimes said to be a bordism is stronger than the homology relation, since homologous submanifolds may turn out not to be bordant.

THEOREM 4 (due to H. Federer [9]). *Let M be a Kählerian manifold, and X a complex submanifold of complex dimension k. Consider all possible real bordism-deformations of X in M. Then Y is an arbitrary bordism-variation of X. The volume of X then does not exceed that of Y if both are compact, i.e., the inequality $\mathrm{vol}_{2k}X \leqslant \mathrm{vol}_{2k}Y$ holds. However, if X and Y are not compact, then a similar inequality occurs for the volumes of those domains in X and Y where they are different. Moreover, if their volumes coincide, then the submanifold is also complex.*

We illustrate by all complex projective subspaces CP^k standardly embedded in CP^n, and globally minimal. Algebraic complex surfaces in C^n or CP^n are also in that class, though possibly possessing singularities.

11. Integral Currents

We have got acquainted with certain concepts of surface and its boundary, starting with the 18–19th centuries ideas to algebraic definitions of a homology boundary. The list of different approaches to such fundamental concepts as surface and its boundary cannot be regarded here as complete. The language invented by H. Federer, W. Fleming, F. Almgren *et al.*, which enabled us to prove many a result briefly discussed in the preceding sections, turned out to be extremely important. This is the language of currents. Here, we only touch upon the concept, referring the reader to [9–11,22,23,25,26,30].

One of the predecessors of the concept of *integral current* is probably the δ-function integration process. If $f(x)$ is a function defined on a domain, then we can consider the function concentrated at a point x_0 from the domain and represent the value $f(x_0)$ as the result of integrating the function $f(x)\delta(x-x_0)$ with respect to the domain $f(x_0)=\int f(x)\delta(x-x_0)dx$. We can also regard the process as integrating the zero-dimensional form on zero-dimensional submanifold x_0 in the domain. Varying x_0, we also alter the result of integration if the function f is not constant. The attempt to extend the definition, and replace the zero-dimensional submanifold with a submanifold of arbitrary dimension k just leads us to the concept of *integral current*. One of the versions of such an extension was suggested by de Rham; however, here, we only give the definition due to W. Fleming and H. Federer. We begin with the concept of smoothable subset. We will say that a certain k-dimensional bounded submanifold X of a Riemannian manifold M is k-smoothable (or k-rectifiable) if X is measurable in the sense of k-dimensional Hausdorff measure [7, 20]; besides, X can be given an arbitrary close approximation by a compact, orientable and smooth k-dimensional submanifold of M. The close approximation concept is sharpened by the fact that for any arbitrarily small positive number ε, there must be such a smooth k-dimensional compact and orientable submanifold $Y \subset M$ that $\mathrm{vol}_k (X \setminus Y) \cup (Y \setminus X) < \varepsilon$, i.e., it is different from X on a set of points of small measure, where vol_k denotes that Hausdorff measure coinciding with the usual k-dimensional Riemannian volume for "nice" sets (submanifolds). We also require that the other inequality $\mathrm{vol}_k (X \cup Y) \setminus (X \cap Y) < \varepsilon$ should hold; however, this does not affect the constructions to follow. It is clear that the stock of k-smoothable subsets is large enough; in particular, it contains all concrete soap films arising in physical experiments. One of the characteristics of the class of k-smoothable sets is that they are most convenient subsets, with respect to which k-dimensional exterior smooth differential forms can be inegrated naturally. Fix k. Let an exterior k-dimensional form ω and a k-dimensional orientable compact submanifold X be given. Then the integral of the form with respect to submanifold $\int_X \omega$ is defined uniquely [1, 5]. This operation can be extended to the class of manifolds X immersed in the ambient manifold M, i.e., self-intersections are possible. Meanwhile, certain sheets of M may be coincident from the point of view of M; however, in integrating ω with respect to X, we naturally take all points of X into account. If the image of X contains parts of several sheets where they coincide, then these portions are taken into account in inegrating with multiplicity. Sheet multiplicity is numbered. We can show that if X is k-smoothable, then, for almost all its points x, there is a unique approximating tangent plane T_x at the point x to X. Since a k-smoothable set can be approximated in an arbitrarily close fashion by a k-dimensional submanifold, then we can extend the integration operation from the class of immersed submanifolds to that of k-*rectifiable subsets*. We omit all technicalities, and assume at once that the integration of any k-dimensional form is defined with respect to any orientable k-rectifiable submanifold X in M.

We obtain a continuous linear functional $\|X\|$ defined on the space $\Lambda^k M$ of all exterior k-dimensional smooth differential forms of M, and mapping the space into R^1. We have to integrate each ω with respect to the k-rectifiable subset X, i.e., $\|X\| : \omega \to \int_X \omega$. $\|X\|$ plays the role of the δ-function in this sense;

however, we consider the k-dimensional subset X in place of one point x_0. As it turns out, $\|X\|$ can be regardel as a "model" of the set X, which is not bad.

We now give a number of important definitions. A current of dimension k in a manifold M is a continuous linear functional mapping the space of k-dimensional exterior forms $\Lambda^k M$ into R. *We call a functional X, where X is a k-rectifiable subset in M, a rectifiable current of dimension k in M.* Let $T: \Lambda^k M \to R$ be a current, and $d: \Lambda^{k-1} M \to \Lambda^k M$ the usual exterior differential operator. The boundary ∂T of a current T is the current $\partial T: \Lambda^{k-1} M \to R$ given by $\partial T(\alpha) = T(d\alpha)$ where $\alpha \in \Lambda^{k-1} M$. The definition generalizes the usual manifold boundary concept. The relation between the definitions is effected via the classical Stokes theorem. In fact, let a current T be rectifiable, i.e., $T = \|X\|$, and X a k-dimensional orientable submanifold with boundary ∂X. Then $\partial \|X\|(\alpha) = \|X\|(d\alpha)$, i.e., $\int\limits_{\partial X} \alpha = \int\limits_{X} d\alpha$. The last equality is called the *Stokes formula* [1, 5]. Thus, if a current is k-rectifiable, and generated by a manifold with boundary, then its boundary is interpreted as a functional generated by the usual boundary of the submanifold X. It goes without saying that, in the general case, this immediate relationship between the current boundary and the set boundary is not there. We now consider the concept of current again.

DEFINITION 1. A current T is said to be *integral* if it is representable as the sum with positive or negative integral coefficients of k-rectifiable currents (if the sum is infinite, we assume that the series converges) and if its boundary ∂T possesses the same property.

Thus, the integral current T is of the form $T = \sum\limits_i a_i \|X_i\|$, where a_i are integers and X_i k-rectifiable subsets of M. The union of the sets X_i is called the *support* of T, i.e., supp $T = \bigcup\limits_i X_i$. The support of a current can be also defined as the least set, outside which the current is annihilated (in integrating forms with respect to sets not affecting the support). The part of integral current is primarily in the possibility to prove existence theorem in their terms for minimal surfaces, while current supports turn out to be smooth almost everywhere, and even analytic submanifolds [9]. Meanwhile, the so-called *mass of a current*, $T = \sum\limits_i a_i \|X_i\|$, is found as usually minimized $m(T) = \sum\limits_i |a_i| \mathrm{vol}_k X_i$.

Many of the above theorems (e.g., the existence of a G-invariant minimal surface) can be formulated and proved in terms of integral currents. Integral currents of dimension k in a manifold M form an additive Abelian group $I_k(M)$. For simplicity, let $M = R^n$. We then write simply I_k. Consider two submanifolds A and B, $A \subset B$, in R^n, and construct the chain complex $\Omega(B, A) = \bigoplus\limits_{k=0}^{n} I_k[B, A]$, where $I_k[B, A] = I_k \cap \{T : \text{supp } T \subset B, \ \text{supp } \partial T \subset A\}$. The boundary operator ∂ is defined by $\partial : I_k[B, A] \to I_{k-1}[B, A]$. Proceeding as in Sec. 1, Ch. I, we can compute the homology $H_i(\Omega(B, A), \partial)$. One important theorem holds good, connecting integral current complex homology with the usual homology of the pair (B, A) with integral coefficients, viz., the isomorphism $H_i(\Omega(B, A), \partial) \overset{\alpha}{\approx} H_i(B, A, \mathbf{Z})$ is valid; in particular, each i-dimensional homology class $\sigma \in H_i(B, A, \mathbf{Z})$ is representable (with the help of α) as an i-dimensional integral current of least mass, compared with those of other currents from the given homology class [9, 10].

12. Spectral Bordisms and the Multidimensional Plateau Problem

Consider a Riemannian manifold M, and distinguish a smooth, compact, and closed (i.e., without boundary) $(k-1)$-dimensional submanifold A, or a "contour". Let there exist at least one k-dimensional manifold W with boundary A. Consider all possible pairs of the form (W, f), where W is a compact manifold with boundary ∂W homeomorphic to A, and $f : W \quad M$ is a continuous (or piecewise smooth) mapping which is identity on ∂W.

PROBLEM A. Can among all pairs of the form (W, f), described above, such a pair (W_0, f_0) be found, that the mapping f_0 or the film $X_0 = f_0(W_0)$, i.e., the image of W_0 in M, may possess reasonable minimality properties; in particular, that $\mathrm{vol}_k X_0 \leqslant \mathrm{vol}_k X$, where $X = f(W)$ is any other film from the indicated class?

Problem A is that of finding the absolute minimum of volume functional for all the films spanning a given boundary frame A (Fig. 63). The following sharpens the problem of the absolute minimum in the class of all bordism-variations of the given submanifold $V \subset M$ (see Sec. 10 of the present chapter).

PROBLEM B. Let (V, g) be a pair of V, a compact closed k-dimensional manifold, and of its continuous (or piecewise smooth) mapping g into M, and $X = g(V)$ the image of V in M. We will say that a pair (V', g') is a bordism-variation of (V, g) if there exists a compact manifold Z with boundary $\partial Z = V \cup (-V')$ and a continuous mapping $F : Z \to M$, so that its restrictions to

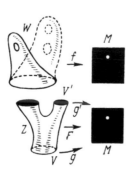

Figure 63

the components of the boundaries of V and V' coincide with the above mappings g and g', respectively. Can, among all bordism-variations, such a pair (V_0, g_0) be found that the image $X_0 = g_0(V_0)$ may obtain reasonable minimality properties, in particular, that $\mathrm{vol}_k X_0 \leqslant \mathrm{vol}_k X$, where X is any other film from the indicated class (Fig. 63)?

In both problems, the topological type of a film can vary, though not quite in an arbitrary manner. In solving the two-dimensional Plateau problem, the following effect is ignored, which starts playing an important role in higher dimensions. Consider the familiar effect of restructuring A catenoid into the union of two disks in extending the boundary frame. The Frame consisting of two circles and spanned by the film $X_t = f_t(W)$ is in Fig. 64. It strives for a position with least area (surface energy). It is clear that, at some moment, the film collapses, and a line segment P joining the upper and lower disks together appears instead of the tube T. We can easily get rid of it in the two-dimensional case, without detriment to the film spanning properties: Its remainder (two disks) still spans the boundary. When $k > 2$ and a $(k-1)$-dimensional submanifold is taken as the boundary, a situation similar to the above occurs, which makes the minimization problem and search for a minimal solution sharply more complicated. As the k-dimensional volume of the deforming film $X_t = f_t(W)$, $0 \leqslant t \leqslant 1$, approaches a minimum, the film tends to take the corresponding minimal position, and gluings similar to the above can arise. In other words, a mapping $f_1 : W \to M$ homotopic to the original can decrease dimension on certain subsets open in W, which leads to pieces P of dimensions less than k in the image $X_1 = f_1(W)$. In contrast to the two-dimensional case, generally speaking, all these pieces cannot be either dropped or mapped continuously into the "massive" k-dimensional part $X^{(k)}$ of the film X_1. Recall that we face the problem of finding a minimal film in the class of those admitting a continuous parametrization via manifolds W; therefore, discarding pieces P of small dimensions in any way, we have to safeguard that the film obtained after the restructuring should still admit a continuous parametrization by means of a manifold. The study in detail of all possible versions of such a restructuring demonstrates that (whether we like it or not) we cannot neglect pieces of small dimensions mechanically arising in minimizing the volume functional, and should take them into account (see the details in [27]).

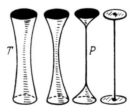

Figure 64

With the purpose of problem simplification, we could confine ourselves only to the k-dimensional volume functional, from whose point all "additions" of small dimensions are inessential, since they have zero k-dimensinal volume. However, it turns out that, even in this case, the solution to the Plateau problem requires that extensive information should be available about the behavior of film pieces of small dimensions [27]. Somehow or other, we again arrive at the problem of studying simultaneously both the functional vol_k and all other functionals vol_s (when s is less k) defined on film pieces of small dimensions. Hence, the mathematical nature of the multidimensional Plateau problem makes us introduce (a) stratified surfaces $X = X^{(k)} \cup X^{(k-1)} \cup \ldots$, where each subset (stratum) $X^{(s)}$ is an s-dimensional "surface" in M, of dimension s at each of its points; (b) stratified volume $SV(X) = (\mathrm{vol}_k X^{(k)}, \mathrm{vol}_{k-1} X^{(k-1)}, \ldots)$ represented by a vector with each component equal to the volume of the corresponding stratum (in each dimension).

Finding a minimal surface X means proof of the existence of a stratified surface whose stratified volume SV is the least. By the minimization of the stratified volume vector, we mean that we first have to minimize its first coordinate (i.e., k-dimensional volume) and then, having fixed this value, to minimize the second component, then, having fixed the minimal values of the first two coordinates, to minimize the third, etc. To realize the program, we need a new language in which the problem can be posed precisely. Minimizing the volume vector is understood lexicographically.

DEFINITION 2. A closed and oriented manifold V^{k-1} is said to be *bordant to zero*, denoted by $V \sim 0$, if there exists a compact and oriented manifold W^k whose boundary ∂W is diffeomorphic to the manifold V, preserving the orientation. Two closed and oriented manifolds V_1 and V_2 are said to be bordant, denoted by $V_1 \sim V_2$, if their disjoint sum $V_1 \cup (-V_2)$ is bordant to zero.

The bordism relation is an equivalence relation. The set Ω_{k-1} of equivalence classes is an Abelian group in which the addition is induced by the manifold disjoint union. The direct sum of the groups $\oplus_k \Omega_{k-1}$ is normally denoted by Ω_*, where the ring structure induced by the direct product of manifolds is defined. Non-orientable bordisms $N_* = \oplus_k N_{k-1}$ are important in the theory of variational problems. To construct the groups N_{k-1}, all closed manifolds are made use of without any restriction on the orientability.

DEFINITION 3. We call the pair (V, f), where V^{k-1} is compact and oriented manifold and f a continuous mapping $V \to M$, an *oriented and singular manifold* of the space M. The set of equivalence classes is endowed with the Abelian group structure, and denoted by $\Omega_{k-1}(M)$. The group is called the $(k-1)$-dimensional oriented bordism group of the space M. A similar construction without regard to the orientability takes us to the definition of non-oriented bordism groups $N_{k-1}(M)$. We now return to the Plateau problem. Problems A and B (see above)' admit the following reformulation.

PROBLEM A. Can, among all compacta X containing A and with the property that the singular bordism (A, i) is equivalent to zero in X, where $i : A \to X$ is an embedding, a compactum X_0 with minimality properties be found?

Since the identity mapping $e : A \to A$ determines an element $\sigma \in \Omega_{k-1}(A)$, the above class of films-compacta X is specified by $i_* \sigma = 0$, where $i_* : \Omega_{k-1}(A) \to \Omega_{k-1}(X)$ is the homomorphism induced by the embedding $i : A \to X$.

PROBLEM B. Can, among all singular manifolds (V, g), $g : V \to M$, bordant (equivalent) to the given manifold (V', g'), $g' : V' \to M$, such a singular manifold (V_0, g_0) be found that the film $X_0 = g_0(V_0)$ may have minimality properties? In other words, can we find such (V_0, g_0) among all representatives (V, g) of the given bordism class $\alpha \in \Omega_k(M)$ that the film $X_0 = g_0(V_0)$ may be minimal in M?

As in Sec. 2, to construct minimization theory accurately, we have to be able to compute bordism groups not only for manifolds and cell complexes, but also for surfaces with singularities, since sufficiently complicated singular points filling up subsets of zero measure are contained in minimal films. This extension of the bordism group concept to a wider class of compact subsets is carried out by means of the so-called spectral process similarly to spectral homology [13–16, 27]. As a result, we obtain groups called *the spectral bordism groups*. An element of such a group is the "limit" of mappings of manifolds. Just as in the case of spectral homology, spectral bordism groups coincide with the groups of usual bordisms if the topological space is "good", e.g., if it is a finite cell complex, or a smooth manifold, or a finite simplicial complex, etc. If the space contains "very complicated points", then the spectral bordism groups may differ from the groups of usual bordisms.

If the boundary A of a film X is a compact smooth manifold, then spectral bordism groups of the boundary A coincide with the groups of usual bordisms of the boundary A. If the film X contains very complicated singularities, then the spectral bordisms of X may a priori differ from usual bordisms. It has not yet been proved that, for minimal surfaces X_0, their spectral bordisms always coincide with usual bordisms (though it seems quite natural). This hypothesis about coincidence of spectral and usual bordisms for globally minimal surfaces was formulated by the author in [27]. We are now prepared to state the theorem solving the Plateau problem in terms of the images of spectra of manifolds with boundary, i.e., when it is posed in a manner close to the classical two-dimensional problem. Fix a "contour" A^{k-1} in R^n, a closed and smooth submanifold taken, for simplicity, as such. Let $A(\sigma)$ be the class of all measurable and compact subsets X of R^n, which contain A and annihilate the bordism $\sigma = (A, e)$, where $e : A \to A$ is the identity mapping. In other words, the singular manifold (A, i), where $i : A \to X$ is an embedding, is null-bordant in the film X, i.e., $i_* \sigma = 0$, where i_* is the homomorphism mapping $(k-1)$-dimensional bordisms of A into $(k-1)$-dimensional spectral bordisms of the film X. Let vol_k be either the Riemannian volume of X or its k-dimensional Hausdorff measure when the film contains sufficiently complicated singularities. The following is due to the author.

THEOREM 5 (see [13–15]). (1) Let $\{X\}_k$ be the class of all compacta X, $A \subset X \subset R^n$, so that $X \in A(\sigma)$ and $\mathrm{vol}_k X = d_k = \inf_Y \mathrm{vol}_k Y$, $Y \in A(\sigma) \backslash (\sigma)$. Then $\{X\}_k$ is non-empty, $d_k > 0$, and each compactum $X \in \{X\}_k$ contains a uniquely determined k-dimensional subset T^k (i.e., of dimension k at each of its ponts), $T^k \subset X \backslash A$, such that $A \cup T^k$ is a compactum in R^n, the set T^k contains a "singular" subset Z_k (which may be empty), where $\mathrm{vol}_k Z_k = 0$ and $T^k \backslash Z_k$ is a smooth minimal k-dimensional submanifold in R^n (in reality, $T^k \backslash Z_k$ is even an analytic submanifold), $T^k \backslash Z_k$ being open and everywhere dense in T^k and $\mathrm{vol}_k T^k = \mathrm{vol}_k X = d_k > 0$.

(2) Furthermore, if $\{X\}_{k-1}$ is the class of all compacta X, $A \subset X \subset R^n$ such that $X \in A(\sigma)$, $X \in \{X\}_k$ and $\mathrm{vol}_{k-1} X \backslash T^k = d_{k-1} = \inf_Y \mathrm{vol}_{k-1} Y \backslash T^k$, $Y \in \{X\}_k$, then the class $\{X\}_{k-1}$ is non-empty; in the case where $d_{k-1} > 0$, each compactum $X \in \{X\}_{k-1}$ contains a uniquely determined $(k-1)$-dimensional subset T^{k-1} (of dimension $k-1$ at each of its points), $T^{k-1} \subset X \backslash (A \cup T^k)$, the subset $A \cup T^k \bigcup T^{k-1}$ being a compactum in R^n, T^{k-1} containing a "singular" subset Z_{k-1} (possibly empty), where $\mathrm{vol}_{k-1} Z_{k-1} = 0$ and $T^{k-1} \backslash Z_{k-1}$ is a smooth and minimal $(k-1)$-dimensional submanifold in R^n, which is everywhere dense in T^{k-1}, with $\mathrm{vol}_{k-1} = \mathrm{vol}_{k-1} X \backslash (A \cup T^k) = d_{k-1} > 0$. However, if $d_{k-1} = 0$, then we put $T^{k-1} = \varnothing$.

(3), (4), ..., etc., with respect to decreasing dimension.

COROLLARY 1. There is always a globally minimal surface $X_0 = A \cup T^k \cup T^{k-1} \cup$ whose stratified volume $SV(X_0) = (\mathrm{vol}_k T^k, \mathrm{vol}_{k-1} T^{k-1}, \ldots)$ is the least (in the above sense) among the stratified volumes of other surfaces $X \in A(\sigma)$. The minimal film X_0 is globally minimal in all dimensions. Each of the strata of X_0 is (except, possibly, a point of set of measure zero) an analytic minimal submanifold in R^n.

A similar theorem also holds in the spectral version of Problem B, i.e., of realizing a spectral bordism with the aid of a film minimal in all dimensions. These theorems are special cases of the general existence theorem for stratified surfaces globally minimal in all dimensins, proved by the author in [13–15] for a wide class of multidimensional variational problems. Thus, in the class of surfaces with fixed boundary A, admitting a continuous parametrization as the spectral images of smooth manifolds with boundary A, there is always a globally minimal surface with the above properties.

13. Existence of a Minimum in Each Homotopy Class of Multivarifolds

Consider a fixed boundary $A \subset M$, where A is a $(k-1)$-dimensional closed manifold. Consider the films obtained from each other by a continuous deformation with boundary A.

PROBLEM A′. Can, among all pairs of the form (W, f), where W is a fixed manifold with boundary A and $f : W \to M$ are all possible continuous (or piecewise smooth) mappings homotopic to a fixed mapping f' and identify on the boundary A (i.e., coinciding with the fixed homomorphism of A onto itself), such a pair (W, f_0) be found that the mapping f_0 or the film $X_0 = f_0(W)$ may obtain the minimality properties; in particular, that $\mathrm{vol}_k X_0 \leqslant \mathrm{vol}_k X$, where $X = f(W)$ is any film from the given homotopy class?

PROBLEM B′. Can, among all mappings $g : V \to M$ (where V is a fixed manifold) homotopic to the original mapping $f : V \to M$, such a mapping g_0 with the minimality property be found, in particular, that $\mathrm{vol}_k g_0(V) \leqslant \mathrm{vol}_k g(V)$, where g is any representative of the above mapping class?

The solution of Problems A′ and B′ does not follow from that of spectral versions of Problems A and B (see above) (the converse is also true). From their standpoint of the ambient manifold M, Problems A and B are those of searching for relative minima in the class of all films, i.e., absolute minima in each homotopy class separately. Dao Chong Thi developed the theory of stratified currents, which is the functional analog of stratified surface theory described above ([13–15]). These stratified currents were called *multivarifolds* in [28].

In [28], the class of locally Lipschitz mappings $g : W^k \to M$ *homotopic to a certain original mapping* $f : W^k \to M^n$ and such that $g|_{\partial V} = f|_{\partial W}$ is considered. A certain multivarifold may be naturally connected with each such mapping ([28]). Since multivarifolds are the functional analogs of stratified surfaces, the Plateau problem may be formulated in the language of multivarifolds *homotopic* to the original, i.e., generated locally by the Lipschitz mapping $f : W^k \to M^n$. Consider the highest-dimensional volume functional (meanwhile, W may be empty). In [28], Dao Chong Thi proved a remarkable theorem whereby *in each homotopy class of multivatifolds there always exists a globally minimal multivarifold, i.e., a multivarifold minimizing the highest dimensional volume functional.*

Many important corollaries follow from this theorem, e.g., the criterion of global minimality was discovered by Dao Chong Thi by means of this theorem ([73, 74]).

14. Cases Where the Dirichlet Problem Has No Solution for the Minimal Surface Equation of Large Codimension

We have got acquainted with the properties of real-valued functions whose graphs are minimal surfaces of codimension one in R^{n+1}. The properties of vector-valued functions are much less known. Their graphs are minimal surfaces of higher codimension. As codimension increases, the properties of graphs turn out to get more complicated, and new effects appear, which were not there for codimension one. We illustrate by a number of examples, mostly on the basis of the work of H. Lawson and R. Osserman ([30]).

Let B be an open subset in $R^n(x_1, \ldots, x_n)$, and $F: B \to R^{n+k}$ a smooth immersion. F determines a minimal surface $F(B)$ in R^{n+k} if and only if F satisfies the differential equation system $\sum_{i,j=1}^{n} \frac{\partial}{\partial x_i}\left(\sqrt{g}\, g^{ij} \frac{\partial F}{\partial x_j}\right) = 0$, where g_{ij} are the components of the metric induced on $F(B)$ by the ambient underlying Euclidean metric, i.e. $g_{ij} = \langle \partial F/\partial x_i, \partial F/\partial x_j \rangle$. Furthermore, g^{ij} are the entries of the matrix inverse to the matrix $\|g_{ij}\|$. Finally, g is the determinant of $\|g_{ij}\|$. The derivation of this vector equation from the Euler equation for the volume functional is left to the reader. We will say that an immersion of F of the domain B in R^{n+k} is non-parametric if F is of the form $F(x) = (x, f(x))$, where f is the mapping (vector-valued function) $f: B \to R^k$. Then the above equation system is of the form

$$(*) \quad \sum_{i=1}^{n} \frac{\partial}{\partial x_i}(\sqrt{g}\, g^{ij}) = 0, \ 1 \leqslant j \leqslant n; \quad \sum_{i,j=1}^{n} \frac{\partial}{\partial x_i}\left(\sqrt{g}\, g^{ij} \frac{\partial f}{\partial x_j}\right) = 0.$$

In other words, $F(B)$ is the graph of $f: B \to R^k$ in R^{n+k}, and the equation system decomposes, where g and g^{ij} are understood in the above sense, and $g_{ij} = \delta_{ij} + \langle \partial f/\partial x_i, \partial f/\partial x_j \rangle$. The obtained system is equivalent to

$$\sum_{i=1}^{n} \frac{\partial}{\partial x_i}(\sqrt{g}\, g^{ij}) = 0, \ 1 \leqslant j \leqslant n; \quad \sum_{i,j=1}^{n} g^{ij} \frac{\partial^2 f}{\partial x_i \partial x_j} = 0.$$

We now formulate the Dirichlet problem for vector-valued functions of the above sort. Let a smooth mapping φ of the boundary ∂B of the domain B in R^k be given. It is required to find a mapping $f: B \to R^k$ such that f is continuous on the closure of B, and locally Lipschitz in B, and f satisfies the minimal surface equation $(*)$. Finally, it is required that the volume of the surface $F(B)$, i.e., the graph of f, should be finite (Fig. 65). We could certainly require that membership of f in a certain wider functional class, and not its smoothness; however, we are not going into the details. We assume B "nice"; e.g., with piecewise smooth boundary. First consider the simplest case where the graph codimension k is one. It is known that the Dirichlet problem for convex domains then has a unique solution for any prescribed continuous boundary conditions. More precisely, the following proposition holds.

PROPOSITION 2. *When $k = 1$, the Dirichlet problem in a convex domain $B \subset R^n$ is always soluble for any continuous initial data, i.e., for any continuous function $\varphi: \partial B \to R^1$. Meanwhile, the solution $f: B \to R^1$ is such that (1) the solution is unique (2) the solution is real analytic and (3) the n-dimensional graph of the*

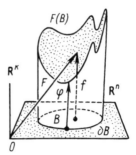

Figure 65

mapping $f : B \rightarrow \mathbf{R}^1$ (i.e., the collection of points of the form $(x, f(x))$ in R^{n+1}, where $x \in B$) is a globally minimal surface in R^{n+1}, i.e., its volume is the least in the class of surfaces with given boundary $(x, \varphi(x))$, where $x \in \partial B \subset \mathbf{R}^n$. The surfaces are meant here in the sense of integral currents.

If codimensions are large, two-dimensional convex domains $B \subset R^2$ should be studied separately; it turns out that the Dirichlet problem always has a solution for them; viz., it is known that, for $n = 2$ and any $k \geqslant 1$ and any convex domain $B \subset R^2$, there is always a solution of the Dirichlet problem (which is continuous on the domain closure) for any non-continuous initial data, i.e., for any continuous mapping $\varphi : \partial B \rightarrow R^k$. However, though the solution existence theorem is valid here, the two-dimensional solution may turn out not to be unique and stable.

As B, consider the two-dimensional disk D^2 of unit radius on the Euclidean plane. Then there exists a real-analytic mapping $\varphi : \partial D^2 \rightarrow \mathbf{R}^2$ of the boundary circle ∂D^2 in $\mathbf{R}^2$ (where $n = 2$ and $k = 1$), so that there are at least three different Dirichlet problem solutions with the same boundary condition φ for the domain D^2. Moreover, one of them (i.e., the graph of the mapping $f : D^2 \rightarrow R^2$) is represented as an unstable minimal two-dimensional surface in four-dimensional Euclidean space, i.e., its area can be decreased with the use of a small continuous deformation fixed on the boundary [30]. To construct an example is a rather delicate matter, and we are forced to omit it because we do not have the space here.

Consider minimal surfaces of greater codimensions $k \geqslant 2$. As the convex domain B in R^n, we take the standard open disk D^n of unit radius. The Dirichlet problem may have no solutions for $n \geqslant 4$. We have established that there exists a mapping $f : \partial D^n \rightarrow \mathbf{R}^k$ for such k, $2 \leqslant k \leqslant n - 2$, that the Dirichlet problem for the minimal surface equation in the ball D^n with boundary conditions f on the sphere $S^{n-1} = \partial D^n$ has no Lipschitz solution F satisfying the condition $F|_{S^{n-1}} = f$. It is curious that the construction of such mappings is based on topological arguments [30].

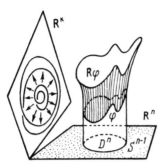

Figure 66

PROPOSITION 3 (Due to H. Lawson and R. Osserman). *Let $n > k \geqslant 2$, $S^{n-1} = \partial D^n \subset \mathbf{R}^n$, $S^{k-1} \subset \mathbf{R}^k$ where S^{n-1} and S^{k-1} are the standard spheres of unit radius. Let $\varphi : S^{n-1} \to S^{k-1}$ be any mapping of class C^2 (i.e., with continuous derivatives until order two inclusive) determining a non-trivial element of the homotopy group $\pi_{n-1}(S^{k-1})$, i.e., non-homotopic to zero. Then there exists a number R_φ such that, for any $R \geqslant R_\varphi$, there exists no Dirichlet problem solution in the ball D^n for the minimal surface equation for codimension k with boundary function $\varphi_R = R \cdot \varphi$ (Fig. 66).*

From the geometric point of view, the following takes place. We consider the graph of the mapping realizing a non-zero element of the homotopy group, and then subject the graph to a simple transformation, i.e., extend it with the similitude coefficient R (Fig. 66). The center is at the origin of coordinates in $\mathbf{R}^k$. Along with extending the boundary data, we can consider the inverse process of contracting the boundary function by multiplying it by decreasing parameter. This process is of interest, because the Dirichlet problem is always soluble for a convex domain, and any codimension for sufficiently small boundary data. In other words, if φ is small in absolute value on ∂B, then there exists a minimal surface close to B and coinciding on the boundary of B with the graph of φ. Hence, multiplying the boundary function by the decreasing parameter r, we finally find the required minimal surface. In other words, there is a value $r_\varphi > 0$ such that, for all $r < r_\varphi$, the Dirichlet problem can be solved in B for the boundary data $\varphi_r = r \cdot \varphi$.

Consider the Hopf mapping studied by us in detail in Ch. 1, taking it as an example. This fibration is given by the mapping $p : S^3 \to S^2$ written in coordinates z, w as $p(z, w) = (|z|^2 - |w|^2, 2zw)$, where z, w are standard Hermitian coordinates in C^2, the sphere S^3 of unit radius is given as usual by $|z|^2 + |w|^2 = 1$, and the sphere S^2 is realized as a unit sphere in $\mathbf{R}^3 = \mathbf{R} \times C$. In fact, it is easy to see by direct calculation that $(|z^2| - |w|^2)^2 + (\operatorname{Re} 2z\bar{w})^2 + (\operatorname{Im} 2z\bar{w})^2 = 1$, i.e., that the vector $p(z, w)$ ranges over the two-dimensional sphere in R^3. The mapping p is known to be non-homotopic to zero, or determines a non-trivial element of the group

$\pi_3(S^2) = \mathbf{Z}$ (see, e.g., [4]). Therefore, by Proposition 3, we obtain that for sufficiently large R, the Dirichlet problem on the four-dimensional ball with boundary conditions $\mathbf{R} \cdot \varphi$ has no solution as the minimal graph of a mapping $F : D^4 \to R^3$. Direct calculation (omitted here) shows that we can take 4.2 for unit radii of the spheres as the boundary value R_φ. Hence, for all $R > 4.2$, the Dirichlet problem with boundary function $R \cdot \varphi$ has no solution [30].

It should be borne in mind that the absence of a solution to the Dirichlet problem with the above restrictions placed does not at all hamper the availability of a four-dimensional minimal surface with a three-dimensional sphere embedded in R^7 as the graph of the mapping $\varphi : S^3 \to S^2 \subset \mathbf{R}^3$ as the boundary "contour frame". Moreover, it follows from the general existence theorems that, for any continuous mapping φ (including those of the above type), there is always a globally minimal surface spanning it in R^7. It follows from Proposition 3 that this surface cannot be represented as the graph of any mapping $F : D^4 \to R^3$.

That a solution of the Dirichlet problem with boundary data $R \cdot \varphi$ exists for sufficiently small R and, on the contrary that it does not exist when the parameter R is large shows that there is a boundary critical value of the parameter R_0, separating these two regions where the multidimensional variational problem is of qualitatively different behavior. It is made clear intuitively that, for $R = R_0$, there must be a singular Dirichlet problem solution for the boundary function $R_0 \cdot \varphi$. In the particular case of a Hopf mapping, considerable symmetries in this fibration enable us to suggest that the singular solution should be represented as a cone whose boundary must be the embedding of the sphere S^3 in R^7, determined by the boundary function $R_0 \cdot \varphi$. This is justified. Thus, in the variational problem in question, minimal cones arise again; we have already encountered them many times above.

Consider a smooth embedding of S^3 in S^6, specified by $i_\alpha(x) = (\alpha x, \sqrt{1 - \alpha^2} \cdot p(x))$, where $p \cdot S^3 \to S^2$ is a Hopf mapping, and $0 < \alpha \leqslant 1, x$ is a point on the unit sphere S^3, the space R^7 is represented as the direct product $R^4 \times R^3$, i.e., $\mathbf{R}^7 = \mathbf{C}^2 \times \mathbf{R}^3$, $\alpha x \in \mathbf{C}^2$, $\sqrt{1 - \alpha^2}\, p(x) \in \mathbf{R}^3$. Since $|x| = 1$ for $x \in S^3$ and $|p(x)| = 1$ for $p(x) \in S^2$, then $|\alpha(x)|^2 + |\sqrt{1 - \alpha^2} \cdot p(x)|^2 = 1$, i.e., $i_\alpha(x) \in S^6 \subset \mathbf{R}^7$. Consider the smooth action of the group $SU(2)$ on $R^7 = C^2 \times R^3$, for which we take the diagonal embedding of $SU(2)$ in the group $SU(2) \times SO(3)$, where each of the factors acts naturally in a manner consistent with the decomposition of R^7 into the direct product. Recall that $SU(2)$, acts on C^2 as the group of special unitary transformations, and as the group $SO(3) = SU(2)/\mathbf{Z}_2$, where Z_2 is the center, on R^3. It is obvious that the Hopf mapping $p : S^3 \to S^2$ is equivariant with respect to these actions of $SU(2)$, hence, in particular, the graph of the mapping is a $SU(2)$-invariant surface in R^7. Since the action of $SU(2)$ is transitive on S^3, the graph is also the orbit of any of its points under the above action of $SU(2)$.

Thus, for any α (where $0 < \alpha \leqslant 1$), the sphere $i_\alpha S^3$ smoothly embedded in the sphere S^6 is the orbit of smooth action of $SU(2)$ on S^6. It is easy to see that it is a generic orbit, or principal orbit (in the sense of Sec. 7 of the present chapter). We know that locally minimal orbits are in a one-to-one correspondence with critical points of the volume functional; in particular, maximum volume orbits are minimal submanifolds in S^6. *Denote by $v(\alpha)$ the volume of the orbit $i_\alpha S^3$.* Straightforward computation shows $v(\alpha) = 2\pi^2 \alpha (4 - 3\alpha^2)^2$ is a function of α, and attains its maximum on the interval $[0, 1]$ for $\alpha = 2/3$. Therefore, $i_{2/3} S^3$ smoothly embedded in S^6 via the mapping $x \to \left(\frac{2}{3} x, \frac{\sqrt{5}}{3} p(x)\right)$, where $x \in S^3$

and $|x| = 1$, is a locally minimal submanifold in S^6, with $\frac{2}{3} x \in \mathbf{R}^4$, $\frac{\sqrt{5}}{3} p(x) \in \mathbf{R}^3$.

Recall that the cone CA, vertex at the point 0, over a smooth submanifold A^q in the sphere S^m is locally minimal if and only if A^q is locally minimal in S^m. Therefore, the four-dimensinal cone $C(i_{2/3} S^3)$, vertex at 0, whose boundary is the submanifold $i_{2/3} S^3$ in S^6 is a minimal surface in R^7, with a singularity at 0. On the other hand, it is obvious that the cone determines a Lipschitz mapping $f_0 : D^4 \to \mathbf{R}^3$, a solution of the Dirichlet problem for the ball D^4 with boundary condition $R_0 \cdot \varphi$ (see above). The statement holds that the Lipschitz mapping

$f_o : \mathbf{R}^4 \to \mathbf{R}^3$ given by the formula $f_0(x) = \frac{\sqrt{5}}{2} |x| p\left(\frac{x}{|x|}\right)$ for $x \neq 0$, where p is a

Hopf mapping, is a solution of the minimal surface equation (∗); in particular, a singular solution of the Dirichlet problem. Lipschitz solutions of system (∗) can be seen not in the smoothness class C^1.

Similar examples are constructable by the aid of two other Hopf fiber bundles, viz., $p' : S^7 \to S^4$ and $p'' : S^{15} \to S^8$. The corresponding orbit volume functions are

$$v'(\alpha) = \frac{\pi^4}{3} \alpha^3 (4 - 3\alpha^2)^2; \quad v''(\alpha) = \frac{2\pi^8}{7!} \alpha^7 (4 - 3\alpha^2)^4.$$

3. Geometry of Volume Functional and Dirichlet Functional Extremals

1. Lower Estimate of the Volumes of Minimal Surfaces

Let X^k be a globally minimal surface of non-trivial topological type in a manifold M^n; e.g., let X realize a cycle from the group $H_k(M, G)$ or a bordism class from the group $\Omega_k(M)$. In studying the properties of such surfaces, the important question of giving a lower estimate of the volume of a surface is often given rise. In other words, what is the least volume of k-dimensional cycles of the manifold M, or the least volume of the supports of these cycles?

Exact estimates of this kind are especially important for the concrete examples of globally minimal surfaces, since, to prove that some or other surface is minimal, we should see that its volume is the lower bound of the volumes of surfaces with the same topological type. To this end, we should be able to calculate this bound efficiently and explicitly. The non-triviallity of the problem is related to the study of large, and not arbitrarily small, variations as is usually done in the proof that a surface is locally minimal. Meanwhile, a lower bound of volumes should be universal, and relate the metric to topological properties of surfaces. It turns out that there is an efficient method for finding the least volumes of closed surfaces of non-trivial topological type in terms of the metric of the ambient manifold [16, 27] (A. T. Fomenko).

Let M^n be a smooth, compact, orientable and connected Riemannian manifold with boundary ∂M, $f(x)$ a smooth function of M, so that it has a unique minimum point x_0 (which is, therefore, absolute), varies from 0 to 1 on M, attains a maximum only on ∂M, being equal to 1 at all its points. Let there be no critical points of f on ∂M, and f the Morse function on the open manifold $M \setminus (\partial M \cup x_0)$ with finitely many critical points. It is clear that, there are no local minima or maxima among them. Consider the level surface $\{f = r\}$, i.e., the set of points x such that $f(x) = r$, $0 \leqslant r \leqslant 1$. We have $\{f = 0\} = x_0$, $\{f = 1\} = \partial M$. Consider a smooth vector field $v = grad\, f$ on M. Let X be an arbitrary k-dimensional measurable and compact subset of M. Consider the "boundary" of this compactum, i.e., the set $\partial X = X \cap \partial M$ (Fig. 67).

Recall the concept of the standard spherical density function $\Psi(x)$ for a subset X of M. Consider a point x. Let B_ε be the n-dimensional ball of geodesic radius ε, center at x. Find the k-dimensional volume (measure) of its intersection with X $vol_k X \cap B_\varepsilon$, and also $\lim\limits_{\varepsilon \to 0} \dfrac{vol_k X \cap B_\varepsilon}{p_\varepsilon} = \Psi(x)$, where $p_\varepsilon = (k$-dimensional volume of standard k-dimensional Euclidean ball of radius ε). $\Psi(x)$ is called the density of X at x. If $x \notin X$, then $\Psi(x) = 0$. If $x \in X$ and x is regular, i.e., X is a smooth submanifold in the vicinity of this point, then $\Psi(x) = 1$, which follows from the fact that, for small ε, $X \cap B_\varepsilon$ is little different

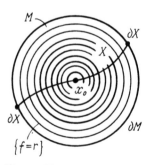

Figure 67

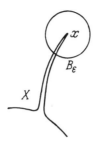

Figure 68

from the plane k-dimensional disk of radius ε. If X is of the form of a sharp "whisker" in the vicinity of x (Fig. 68), then the k-dimensional volume of the intersection of X with B_ε is small; therefore $\Psi(x) < 1$. Otherwise, if $x \in X$ is singular (e.g., a number of k-dimensional sheets of X converges at it as in the case of two-dimensional soap films), then $\Psi(x) > 1$. Thus, the density function reflects the local properties of X sufficiently well.

Consider the variational problem in the class of surfaces spanning a fixed contour A, where A is a $(k-1)$-dimensional measurable compact subset of ∂M, so that $H_{k-1}(A, G) = 0$. We introduce the class of surfaces for which we shall solve the problem of absolute minimum existence. Consider the class A^* of all copacta X with boundary $A \subset \partial M$, so that (1) $\dim X = k$, and the compactum X is measurable, (2) X does not contain any "whiskers", i.e., for any point $x \in X \setminus \partial X$, we have $\Psi(x) \geqslant 1$; (3) the boundary of X contains A, i.e., $A \subset \partial X$, and (4) in embedding A in X, the whole $(k-1)$-dimensional homology group of A is annihilated, i.e., the homomorphism $i : H_{k-1}(A, G) \to H_{k-1}(X, G)$ induced by the embedding $i : A \to X$ is trivial, where H_{k-1} are spectral homology groups (see above).

If d is the infimum of the volumes of all surfaces from A^*, then there exists a globally minimal surface $X_0 \in A^*$ such that $\mathrm{vol}_k X_0 = d$. All these and results to follow of the present section are also valid for the cohomology (including the generalized); however, we are not going to dwell on this here [16, 27]. For X_0, we also have $\Psi_k(x) \geqslant 1$ for any $x \in X_0 \setminus \partial X_0$. We assume that X_0 passes through a point x_0, of absolute minimum for the function f. We also construct the important function $\psi(r) = \mathrm{vol}_k X_0 \cap \{f \leqslant r\}$ called the surface volume function. By $\{f \leqslant r\}$, we denote the n-dimensional subset of M, consisting of points x such that $f(x) \leqslant r$ (Fig. 69). Generally speaking, the volume function is different for different X_0, and defined on $[0, 1]$, obviously not decreasing as r increases from zero to 1. The problem described at the beginning of the section can now be formulated as that of finding an exact lower estimate on ψ in terms of the metric of M irrespective of the topological type of the minimal surface X_0. The exactness is meant in the sense that the required estimate should turn into equality for sufficiently pithy series of concrete triples (M, f, X).

It is clear that $\psi(1) = \mathrm{vol}_k X$, since $X \cap \{f \leqslant 1\} = X \cap M = X$. Thus, knowing

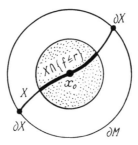

Figure 69

the lower estimate on $\psi(r)$, we shall be able also to give a lower estimate of $\psi(1)$, or the minimal surface volume. Meanwhile, we can vary the boundary of A in ∂M, only seeing to the non-triviality of $H_{k-1}(A, G)$ *and annihilation by the homomorphism* i_*.

2. Vector Field Deformation Coefficient

We now introduce the concept of vector field grad f deformation coefficient. Owing to the property of the function f, almost all integral paths of the field start at x_0, and end on the boundary ∂M. We assume that a point $x \in M$ is said to be *generic* if an integral path starting at x_0 and ending on ∂M passes through it. Consider a small $(k-1)$-dimensional ball D_ε of radius ε, orthogonal to the vector grad $f(x)$, in the manifold M at x (Fig. 70), and the set of all integral paths from x_0 to the points of D_ε. Denote the set of points swept out by the paths by CD_ε. From the point of view of geometry, this k-dimensional set is a "cone", vertex at x_0 and D_ε as the base. The cone is obtained if we start deforming D_ε, shifting it along the integral paths of grad f towards x_0. The ball then sweeps out CD_ε. If x is generic, then the cone is a smooth k-dimensional submanifold with foundary on M. If x is arbitrary, then the cone is defined

Figure 70

similarly; however, it is, generally speaking, a cell complex, and not a submanifold.

We now find the value $\lambda(x) = \lim\limits_{\varepsilon \to 0} \dfrac{\operatorname{vol}_k CD_\varepsilon}{\operatorname{vol}_{k-1} D_\varepsilon}$. In other words, we are looking for the ratio of the cone volume to the volume of its base, and pass to the limit when the base is contracted to a point. It is obvious that $\lambda(x)$ depends on the choice of a $(k-1)$-dimensional disk D_ε orthogonal to the vector $\operatorname{grad} f(x)$. Consider the set of all such disks. Let $\kappa_k(x) = \sup \lambda(x)$, where sup is found for all D_ε. The function $\kappa_k(x)$ is smooth on M. We call it the *k-dimensional vector field grad f deformation coefficient*. Generally speaking, it is different for different x, since it is closely related to the Riemannian curvature tensor of M. $\kappa_k(x)$ can be calculated easily in many cases. We illustrate by example. Let M be a ball of radius R in R^n, and $f(x) = |x|$ the distance from the point $0 = x_0$ to a point x, $v = \operatorname{grad} f$. Then $\kappa_k(x) = r/k$, where $r = |x| = f(x)$.

3. Closed Surfaces of Non-Trivial Topological Type and Least Volume

The above coefficient is important in computing the least volume of globally minimal surfaces (see the works of the author).

THEOREM 1 [16]. *Let f be a smooth function on M^n, so that $0 \leqslant f \leqslant 1$, $f(x_0) = 0$, $f(\partial M) = 1$, and a Morse function on $M \setminus (x_0 \cup \partial M)$ with only saddle critical points whose indices do not exceed $k - 2$, when k is an integer, $k < n$. Suppose that X_0^k is a globally minimal surface in the above class, passing through x_0. The inequality $\psi(r) \geqslant q(r) \cdot l$ then holds for this surface volume function, where the constant $l = \lim\limits_{a \to 0} \dfrac{\psi(a)}{q(a)}$ does not depend on the parameter r and is only determined by the structure of the minimal surface X_0 around x_0, whereas the function $q(r)$ is of the form*

$$q(r) = \exp \int^r [\max_{x \in \{f = r\}} (\kappa_k(x) \cdot |\operatorname{grad} f(x)|)]^{-1} dr.$$

In particular, $\psi(1) = \operatorname{vol}_k X \geqslant l \cdot q(1)$. Thus, the behavior of the volume function $\psi(r)$ on X_0 is determined by that at the initial moment $r = 0$, i.e., in the neighborhood of x_0 and by the ambient manifold geometry. This estimate is exact in the sense that there are rich series of the triples (M, f, X) for which the inequality turns into equality, in which cases we obtain a precise expression both for the volume function and minimal surface volume.

Thus, to obtain an estimate for surface volume, we first should calculate the absolute value of the gradient of the function f on the level surface $\{f = r\}$ with respect to the metric given on M, multiply it by the deformation coefficient of the field grad f, find the maximum α of this product on the level surface, and integrate the function $1/\alpha$ relative to r. The function $q(r)$ specified by the indefinite integral is certainly determined up to constant multiplier; however, this does not affect the estimation, for $1/q(a)$ is involved in the formula for $l1$, which removes indeterminacy in multiplying $q(r)$ by $l1$.

Theorem 1 follows from one general statement about the behavior of the normalized volume function. Let a triple (M, f, X) satisfy the conditions of the theorem. Then the piecewise continuous $\psi(r)/q(r)$ does not decrease with respect to r as r increases from 0 to 1. Hence, $\dfrac{\psi(1)}{q(1)} = \dfrac{\text{vol}_k X_0}{q(1)} \geqslant \lim\limits_{a \to 0} \dfrac{\psi(a)}{q(a)} = l$, i.e., $\text{vol}_k X_0 \geqslant l \cdot q(1)$. A similar statement also holds for closed minimal surfaces realizing non-trivial (co)cycles in the ambient manifold M [16].

We now consider some corollaries to Theorem 1; e.g., the problem of giving a lower estimate for the volume of the minimal surface X^k passing through the center of the standard Euclidean ball D^n in R^n. Plane k-dimensional disks, i.e., sections of D^n by k-dimensional planes passing through the center, can be naturally distinguished from this class. The volume of the part of the surface inside the ball is known to be given a lower estimate by the volume of the plane k-dimensional disk in many cases; e.g., where the surface X^k is complex analytic in $\mathbf{R}^{2n} = \mathbf{C}^n$. It turns out that one general statement valid for any globally minimal surfaces (corresponding to any extraordinary homology or cohomology theory of coefficients) also holds.

PROPOSITION 1. *Let X^k be the globally minimal surface in $\mathbf{R}^n$, passing through the center of the ball D^n of radius r, with boundary on the ball boundary (in the sense of A^*, see above). Then $\text{vol}_k X \cap D^n \geqslant \Psi(0) r^k \gamma_k \geqslant r^k \gamma_k = \text{vol}_k D^k$, where $\Psi(0)$ is the density of the surface X at the point 0, γ_k the volume of the unit Euclidean ball, and $r^k \gamma_k$ the volume of the k-dimensional Euclidean ball of radius r. Besides, $\Psi(0) \geqslant 1$; equality occurs if and only if 0 is regular on X^k. Thus, the volume of any globally minimal surface of any codimension, passing through the center of the Euclidean ball, with the boundary on the ball boundary is not less than the volume of the standard plane section of the ball by a plane, i.e., the volume of the plane k-dimensional disk.*

Let M^n be a closed, connected and compact Riemannian manifold, and x_0 its point. It is known that M is representable as a cell complex containing only one n-dimensional cell homeomorphic to the disk D^n whose boundary is glued to a subcomplex C, $\dim C \leqslant n - 1$. Then the set $M \backslash (C \cup x_0)$ is homeomorphic to the open disk D^n without x_0. Consider the function f on M, which is smooth on $M \backslash (C \cup x_0)$, with precisely one minimum on M at x_0 (assuming that $f(x_0) = 0$), taking a maximum value 1 on C, while without critical points on the open manifold $M \backslash (C \bigcup x_0)$. Denote the set of all such functions by $F(x_0)$. Considering the field grad f, where $f \in F(x_0)$ on $D^n \backslash x_0$ proceeding as above, we can define the function

$$q(r) = \exp \int \frac{dr}{\max\limits_{x \in \{f = r\}} (\kappa_k |\operatorname{grad} f|)}$$

depending on x_0 and the choice of f.

DEFINITION 1. *We call the function* $\gamma_k q(1) \lim\limits_{a \to 0} \dfrac{s^k}{q(a)} = \Omega(x_0, f)$, *where* $q(1) = \lim\limits_{r \to 1} q(r)$, *the manifold* M *k-dimensional F-nullity function.* Consider the value $\Omega_k(x_0) = \sup\limits_{f \in F(x_0)} \Omega(x, f)$ at each point $x_0 \in M$. Finally, we call the value $\Omega_k = \inf\limits_{x_0 \in M} \Omega_k(x_0)$ the k-dimensional F-nullity of the manifold M.

This depends on the dimension k and M. It is clear that the inequality $\Omega_k > 0$ always holds. Consider the (co)homology $H_k^{(k)}(M, G)$ with coefficients in a group G. Let L be an arbitrary non-trivial subgroup of $H_k^{(k)}(M, G)$. Suppose that X^k is a globally minimal surface in M, which realizes the subgroup, which means that, under the embedding $i : X \to M$, the surface X realizes all (co)cycles from L. In the homology case, the condition $L \subset \operatorname{Im} i_*$ should hold, where $i_* : H_k(X) \to H_k(M)$; in the cohomology case, $L \subset H^k(M) \backslash \operatorname{Ker} i^*$, where $i^* : H^k(M) \to H^k(X)$. Let $x_0 \in X$, and $\Psi(x_0)$ the value of the density function for X at this point.

THEOREM 2 (due to A. T. Fomenko [16]). *Let* $X^k \subset M^n$ *be a closed globally minimal surface in* M, *realizing a non-trivial subgroup* L. *Then* $\operatorname{vol}_k X \geqslant \Psi(x_0) \cdot \Omega_k \geqslant \Omega_k > 0$. *The value* Ω_k *thereby turns out to be a universal constant supplying a lower estimate for the k-dimensional volume of any closed minimal surface realizing a non-trivial (co)cycle in the manifold* M. *There are pithy series of examples where the estimate is attained on concrete minimal surfaces of non-trivial topological type, i.e., the inequality turns into equality; in particular, if, for the minimal surface* X, $\operatorname{vol}_k X = \Omega_k$, *then the surface is a smooth, minimal, compact and a closed submanifold of* M.

This theorem due to the author enables us to prove that many concrete surfaces are minimal; e.g., if M is the real projective space with standard metric, then all projected subspaces standardly embedded in M are globally minimal surfaces realizing non-trivial (co)cycles. A similar statement also holds for complex and quaternion subspaces in complex and quaternion projective spaces, respectively. The sphere S^8 in the homogeneous space $M^{16} = F_4/\operatorname{Spin}(9)$ (meaning the standard embedding) and standardly embedded subgroup SU_2 in any compact Lie group are a globally minimal surfaces realizing a non-trivial (co)cycles. We have: $\operatorname{vol}_k X = \Omega_k$ for all these surfaces. They are totally geodesic submanifolds, and not only minimal. These surfaces are of least volume among all those realizing non-trivial (co)cycles in the given dimension. If X is any closed quaternion submanifold of real dimension 4 in Kählerian, quaternion, symmetric, connected and compact space, then it is a globally minimal surface in its (co)homology class. Complex, quaternion and real Grassmann submanifolds standardly embedded in Grassmann manifolds of the same type are also globally minimal surfaces realizing non-trivial (co)cycles and with the property $\operatorname{vol}_k X = \Omega_k$, with certain restrictions placed on the dimension in the real case (see for other examples in [15, 16, 27, 33]).

4. The Cases Where There Are No Local Minima of the Dirichlet Functional in the Class of Mappings of Homogeneous Spaces to an Arbitrary Riemannian Manifold.

Let M^m and N^n be two smooth Riemannian manifolds without boundary. M is compact, closed and orientable. We shall be interested in harmonic mappings $f : M \to N$. A certain value $D[f]$ is naturally related to each smooth mapping $f : M \to N$, which enables us to define the multidimensional Dirichlet functional. Let g_{ij} and $\hat{g}_{\alpha\beta}$ be two Riemannian metric on M and N, respectively. The differential df is naturally related to the mapping f, mapping the tangent space $T_x M$ into $T_{f(x)} N$. Let $x_1, \ldots, x_m$ be local coordinates on M^m, and $y_1, \ldots, y_n$ be those on N^n. Then the mapping f is written as a set of functions $y_\alpha = y_\alpha(x_1, \ldots, x_m)$, $1 \leqslant \alpha \leqslant n$. Let g^{ij} be the constituents of the matrix inverse to the matrix $\|g_{ij}\|$, i.e., $\sum_i g^{ij} g_{iq} = \delta_q^j$, and $d^m x$ the standard n-dimensional exterior Riemannian volume form on M. We define the Dirichlet functional by

$$D[f] = \int_M \sum_{i,j,\alpha,\beta} g^{ij} \frac{\partial y_\alpha}{\partial x_i} \frac{\partial y_\beta}{\partial x_j} \hat{g}_{\alpha\beta} d^m x.$$ Thus, in computing the Dirichlet integral,

both fundamental tensors are involved. That M is compact and orientable is assumed for the definition to be correct. The Dirichlet functional is defined on the infinite-dimensional space of smooth (piecewise smooth) mappings of M to N. Sometimes this functional is also said to be energy functional. Harmonic mappings are its extremals. In the special case where M is a two-dimensional surface and $N = R^3$, we obtain the harmonic surfaces studied in Sec. 1, Ch. 4, 12. The energy functional properties and its extremals are discussed, e.g., in [27]. We denote by $\delta_\eta D[f]$ the first variation (derivative) of the functional D at a point f along the vector field η which is a field on the manifold N, and defined in the neighborhood of the image $f(M)$ in N. Fields of a certain class are usually considered; e.g., H_1^2 [1]. Let f_0 be a critical point (extremal) of the Dirichlet functional, i.e., $\delta_\eta D[f] \equiv 0$, for any perturbation of η at f_0 and $\delta^2 D$ the second variation of Dirichlet functional at f_0. It is convenient to represent the latter as a bilinear form on the linear space of all vector fields defined in a small neighborhood of $f(M)$ in M. Consider that maximum subspace of vector fields, on which the form $\delta^2 D$ is negative definite. The vector fields from this subspace are characterized by a decrease in the functional value for a perturbation of the Dirichlet functional along them. The subspace dimension is called the index of the critical point f_0, i.e., *of the harmonic mapping* f_0. We denote it by ind f_0. If the index of a harmonic mapping is zero, then the mapping is a local minimum for the Dirichlet functional. If the index is positive, then the extremal f_0 is a saddle point, and there are perturbation directions along which energy decreases. It is known that the index of a harmonic mapping is always finite [37]. To show that the Dirichlet functional has no local minima in the class of mappings $f : M \to N$, it suffices to prove that the indices of all harmonic mappings from the class are strictly positive. Many examples of manifolds M and N turn out to exist with the Dirichlet functional having no local minima in the class of $f : M \to N$ other than trivial, or those mappings for which M is sent into a point. For description, recall a number of definitions.

DEFINITION 2. A compact, irreducible and homogeneous space (manifold) M is the homogeneous space $M = G/H$ obtained by taking the quotient of the compact Lie group G relative to the compact subgroup H, the linear representation of H on the tangent space to M at the point $x_0 = eH$ being irreducible, i.e., it has no non-trivial invariant subspaces.

The space M is identified with the set of left cosets gH relative to the subgroup H, and is the stability sub-group of the point $x_0 = eH$, i.e., leaves it fixed under the natural right action of H on M when $gH \to hgH$, $g \in G$, $h \in H$. Since x_0 is fixed, the tangent space to M is mapped into itself by the differential of each transformation from H. This is just what specifies a linear representation of H on $T_{x_0} M$. Meanwhile, we assume that the bi-invariant Riemannian metric is specified on the group G (existing on any compact Lie group), naturally inducing a certain metric on M, invariant under the action of G. Therefore, G acts on M via isometries; e.g., an irreducible, compact and homogeneous space is given by the sphere S^{n-1} with the standard metric, admitting a representation in the form $SO(n)/SO(n-1)$, where the subgroup $SO(n-1)$ is standardly embedded in $SO(n)$.

Let $M = N$, and $f : M \to N$ be the identity mapping. It is easy to see that it is harmonic. Therefore, the index of the identity mapping of M onto itself is defined. We now formulate the results obtained by A. I. Pluzhnikov [90]–[95].

THEOREM 3. *Let M be a compact, irreducible and homogeneous Riemannian space, and $f : M \to N$ a non-compact (different from a mapping to a point) and harmonic mapping of M into any smooth Riemannian manifold M. Assume that the identity mapping index of M mapped onto itself is positive. Then the index of $f : M \to N$ is also positive. In other words, there are no Dirichlet functional local minima among all harmonic mappings of M into N different from a mapping into a point.*

Before illustrating such manifolds, we see that the results of this sort point out to profound difference between the Dirichlet functional and the surface volume functional. We know that, in each non-trivial homotopy class of a multivarifolds, corresponding to a fixed mapping $f : M \to N$, there is a multivarifold, corresponding to a mapping f_0 homotopic to the original, on which the volume functional attains the absolute minimum (see Sec. 2, Ch. 4). Moreover, there is always an absolute minimum in any non-trivial spectral bordism class (Sec. 2, Ch. 4). In contrast to these results, we see that there are closed manifolds M for which there are no local minima in the class of mappings $f : M \to N$ homotopic to a non-constant mapping, i.e., the Dirichlet functional is not only without local minima but also does not attain an absolute minimum on the class of homotopy non-trivial (non-contractible) mappings, where we deal with a closed manifold M. However, if we consider manifolds with boundary, then there are cases where the Dirichlet functional attains an absolute minimum in the class of homotopy non-trivial mappings (see e.g., [27]). We illustrate by examples of manifolds satisfying the conditions of the theorem.

PROPOSITION 2. *No non-constant harmonic mapping of the following compact, and closed Riemannian manifolds M into any Riemannian manifold N is not a Dirichlet functional local minimum: (1) standard spheres S^m for $m \geqslant 3$, (2) quaternion Grassmann manifolds $\mathrm{Sp}(p + q)/(\mathrm{Sp}(p) \times \mathrm{Sp}(q))$, and (3) manifolds $SU(2p)/\mathrm{Sp}(p)$ for $p > 1$.*

All these manifolds are compact, irreducible and homogeneous spaces for which the identity mapping index is positive [38]. If the manifold M is a sphere, then the statement of Proposition 2 can be strengthened. First, a stronger estimate for an arbitrary harmonic mapping index can be obtained, and, second, the Dirichlet functional unattainable absolute minimum be shown to be equal to zero.

PROPOSITION 3. *Let $f : S^m \to N$ be a non-constant harmonic mapping of the standard sphere, where $m \geqslant 3$, into an arbitrary smooth Riemannian manifold N. Then the inequality ind $f \leqslant m + 1$ holds.*

This estimate is, generally speaking, cannot be improved, since it is shown in [38] that, in the particular case of the identity mapping of the sphere onto itself for $m > 3$, its index is precisely $m + 1$.

PROPOSITION 4. *The Dirichlet functional absolute minimum on any connected component of the space $C^\infty (S^m, N)$ of smooth mappings of the standard sphere S^m for $m \geqslant 3$ to an arbitrary Riemannian manifolds N is zero. In other words, in any homotopy class of mappings of the sphere S^m to N, there are always mappings with arbitrarily small positive energy.*

Moreover, if the manifold M is such that the Dirichlet functional absolute minimum vanishes on the class of all mappings homotopic to the identity mapping of M onto itself, then the Dirichlet functional minimum is zero also on any homotopy class of mappings of M to an arbitrary smooth Riemannian manifold N.

5. Cases Where There Are No Dirichlet Funcational Local Minima in the Class of Mappings of an Arbitrary Riemannian Manifold to a Homogeneous Space.

It turns out that there are circumstances in a certain case dual to the results of the previous section. Consider smooth mappings of an arbitrary smooth Riemannian manifold M to a compact, irreducible and homogeneous space N. In other words, we interchange M and N from the preceding section, though preserving the prior notation for the image and inverse image. We illustrate by the results obtained by A. V. Tyrin, [86–89].

THEOREM 4. *Let M be a connected, compact, orientable, closed and smooth Riemannian manifold, and N a compact, irreducible and homogeneous space supplied with the natural, invariant and Riemannian metric. Assume that the index of the identity mapping of N onto itself is positive. Then any harmonic mapping $f : M \to N$ which is not constant (i.e., other than a mapping into a point) also has a positive index, and is not a local minimum point for the Dirichlet functional.*

Thus, as in Sec. 4, the problem of positivity of index of the identity mapping of an irreducible homogeneous space onto itself is central in solving the problem of existence of the energy functional local minima on mapping $f : M \to N$ spaces. The calculation of the identity mapping index for many cases was carried out by R. Smith in [38].

PROPOSITION 5 [38]. *Let $N = G/H$ be a compact, irreducible and symmetric space, where G is one of the classical Lie groups (i.e. G is not a special group). Then the index of the identity mapping of N onto itself is always zero except when*
  (1) *The sphere S^m for $m \geqslant 3$,*
  (2) $Sp(p+q)/(Sp(p) \times Sp(q))$,
  (3) $SU(2p)/Sp(p)$ *for $p \geqslant 1$.*

Combining this result with Theorem 4, we obtain the statement.

PROPOSITION 6. *Of all classical, compact, irreducible and symmetric space $N = G/H$, the spaces S^n, $n \geqslant 3$; $Sp(p+q)/(Sp(p) \times Sp(q))$; $SU(2p)/Sp(p)$, $p > 1$ and they only, fulfill the property that no smooth non-constant mapping $f : M \to N$ of any connected, compact and orientable Riemannian manifold M to a manifold N is not a local minimum point for the Dirichlet functional; in particular, the energy absolute minimum is not attained either.*

For all the above manifolds N, we can indicate that class of deformations among which, for any harmonic mappings $f : M \to N$, there is necessarily a deformation increasing the Dirichlet functional value on the mapping. Generally speaking, for any harmonic mapping f, there is its own decreasing deformation from the class.

Consider the particular case of $N = S^n$, $n \geqslant 3$. We can then exhibit explicitly the class of deformations containing the decreasing deformation which decreases the value of the Dirichlet functional for the harmonic mapping $f : M \to S^n$, i.e., making the zero contribution to the index of the critical point f, for which it suffices to consider the restrictions to the sphere S^n (standardly embedded in R^{n+1}) of coordinate linear functions generated by Cartesian coordinates in R^{n+1}. In other words, consider the functions $\lambda_i(x) = x_i$, where $1 \leqslant i \leqslant n + 1$, x_i is the i-th Cartesian coordinate of the point x, on the sphere S^n. Finally, take their gradients. We then obtain a set of $n + 1$ vector fields on the sphere. These are just what determines the class of deformations containing the one decreasing the Dirichlet functional at a critical point which is the identity mapping.

A two-dimensional sphere is not involved in the above results. As a matter of fact, the statements of Theorems 3 and 4 do not hold for this manifold. Any holomorphic mapping of Kählerian manifolds is known to realize the Dirichlet functional minimum in its homotopy class [37]. Therefore, in particular, the identity mapping of S^2 onto itself is harmonic, and realizes the Dirichlet functional minimum in its homotopy class (e.g., see the elementary proof in [1]).

A series of harmonic mappings of spheres $S^m \to S^n$ was constructed in [39]; however, the question whether or not they are Dirichlet functional local minima remained open. It follows from Proposition 6 that these mappings are not energy minima for $n \geqslant 3$. It would be interesting to extend the computation of identity mapping indices for compact and irreducible spaces different from those considered in [38]. We now study the case where a simple and compact Lie group is such, i.e., the classical group $A_l = SU(l + 1)$, $l \geqslant 1$; $B_i = SO(2l + 1)$, $l \geqslant 2$; $C_l = Sp(2l)$, $l \geqslant 2$ or $D_i = SO(2l)$, $l \geqslant 3$, or one of the special groups G_2, F_4, E_6, E_7, E_8 (due to A. V. Tyrin).

PROPOSITION 7. *Let $N = G$ be a simple and compact Lie group $A_l, l \geqslant 1; B_l, l \geqslant 2; C_l, l \geqslant 2; D_l, l \geqslant 3; G_2, F_4, E_6, E_7, E_8$ with the bi-invariant metric. Then the index of of the identity mapping of N onto itself is zero except the cases of (1) C_1 for $1 \geqslant 2$, (2) D_3, (3) A_1, $1 \geqslant 1$.*

Making use of Theorem 4, we obtain the following statement:

PROPOSITION 8. *Of all simple, simply connected and compact Lie groups $N = G$ for $SU(l + 1)$, $l \geqslant 1$; $Sp(2l)$, $l \geqslant 2$ and $Spin$ (6), and only they, no smooth non-constant mapping $f : M \to N$ of any connected, compact, and orientable Riemannian manifold M to the group N is a local minimum point for the Dirichlet functional.*

The above results regard homogeneous spaces endowed with natural invariant metrics. The question arises regarding their stability under small perturbations of Riemannian metrics. It turns out that the stability is there in certain cases (A. V. Tyrin).

Let N^n be a compact, closed and smooth manifold in the sphere S^n, N belonging to no equator S^{p-1} obtained by the section of the sphere S^p by a hyperplane passing through the center. Suppose that B is the second fundamental form of embedding for the submanifold $N \subset S^p$ (e.g., see the definition and properties in [1, 27]). This form $B(X, Y)$ is defined on pairs of vectors X, Y tangent to N. Let $e_1, \ldots, e_n$ be the orthonormal basis in the tangent plane to N. Then we define the norm $\|B\|$ of B by

$$\|B\|^2 = \sum_{i,j=1}^{n} |B(e_i, e_j)|^2, \quad |B(X, Y)| \text{ denotes the length of the vector } B(X, Y).$$

Recall that B assumes values in the set of vectors orthogonal to N.

THEOREM 5. Let N^n be a compact, smooth and closed submanifold in the standard sphere S^p and B the second fundamental form of embedding N in S^p. Assume that the embedding $N \subset S^p$ satisfies the condition

$$\left(1 + \max\left\{1, \frac{\sqrt{n-1}}{2}\right\}\right) \cdot \|B\|^2 < n - 2,$$

where $n = \dim N \geqslant 3$. Then, for any smooth non-constant harmonic mapping f of any smooth, connected, orientable, compact and closed manifold M to the manifold N, the mapping index is positive; in particular, no constant harmonic mapping $f : M \to N$ is a local minimum point for the Dirichlet functional, and it does not attain an absolute minimum on any non-trivial homotopy mapping class $f : M \to N$.

Thus, Proposition 6 holding for the standard sphere S^n, $n \geqslant 3$, remains valid also for all Riemannian metrics on it, obtained by a small perturbation of the standard one. The condition of Theorem 5 means that the submanifold N is not very much "curved" in S^p, since the restriction to the second form norm means that N does not make any sharp local rotations in the direction of its normal planes. If N is a section of the sphere by a plane, then the form B vanishes, and the condition of Theorem 5 does hold. The stock of manifolds N for which the inequality holds is not large. Such a manifold is a homotopy sphere; therefore, it is homeomorphic to the standard one for $n \geqslant 4$. Hence, Theorem 5 can be regarded as the stability theorem for the statement about the absence of energy local minima in the class of mappings $f : M \to S^n$ for a sufficiently small perturbation of the metric on the standard sphere S^n.

In conclusion, we give one theorem due to A. I. Pluzhnikov [95].

THEOREM 6. Let M be a smooth, connected, closed and orientable manifold. Then the following statements are equivalent.

(1) M is two-connected, i.e., its first homotopy groups vanich $(\pi_1(M) = \pi_2(M) = 0)$.

(2) At least in one Riemannian metric on M, the energy functional minimum is zero on the homotopy class of the identity mapping of M into itself.

(3) For any smooth Riemannian manifold M and an arbitrary choice of the Riemannian metric on M, the energy functional minimum is zero on all connected components of the mapping space $C^\infty(N, M)$.

(4) For any smooth, compact and orientable Riemannian manifold N and an arbitrary choice of a Riemannian metric on M, the energy functional minimum is zero on all connected components of the mapping space $C^\infty(M, N)$.

(5) For any smooth, compact and orientable Riemannian manifold N and any Riemannian metric on M, the energy functional global minimum is only attained in one homotopy class of mappings from N to M, viz., on locally constant mappings.

The condition for two-connectedness of the manifold, which is equivalent to Conditions 2–5, is what is essentially new. Other statements are based on the well-known results obtained by J. Sacks, K. Uhlenbeck, J. Eels, R. Smith, J. Sampson, and P. Leung.

The generalization of Theorem 6 was obtained by B. White in his interesting works [112] and [129], of which the latter contains the general theorem describing the conditions of existence of energy minimizing maps.

References

1. Dubrovin, B.A., Fomenko, A.T. and Novikov, S.P., Modern Geometry. Methods and Applications. Springer-Verlag. GTM 93, Part 1, 1984; GTM 104, Part 2, 1985.
2. Milnor, J., Morse Theory. Princeton, N.J., 1963.
3. Milnor, J., Lectures on the h-cobordism theorem. Princeton University Press, Princeton, N.J., 1965.
4. Fomenko, A.T., Fuchs, D.B. and Gutenmacher, V.L., Homotopic Topology. Akadémiai Kiadó. Budapest, Hungary, 1986.
5. Mishchenko, A. S. and Fomenko, A.T., A Course of Differential Geometry and Topology. Moscow University Press, Moscow, 1980 (Russian).
6. Raschevsky, P. K., Riemannian Geometry and Tensor Analysis. Nauka, Moscow, 1967 (Russian).
7. Reifenberg, E.R., Solution of the Plateau problem for m-dimensional surfaces of varying topological type. Acta Math., V. 104, 1 (1960), pp. 1–92.
8. Morrey, C.B., Multiple integrals in the calculus of variations. Berlin, Springer, Bd. 130, 1966.
9. Federer, H., Geometric measure theory. Springer, Berlin, Bd. 153, 1969.
10. Almgren, F.J., Existence and regularity almost everywhere of solutions to elliptic variational problem among surfaces of varying topological type and singularity structure. Ann. of Math., Ser. 2, V. 87, 2 (1968), pp. 321–391.

11. Almgren, F. J. and Taylor, J., The geometry of soap films and soap bubbles. *Scientific American*, July 1976, pp. 82–93.

12. Fomenko, A.T., Multidimensional variational methods in the topology of extremals. *Russian Math. Survey*, **V. 36, 6** (1981), pp. 127–165.

13. Fomenko, A.T., The multidimensional Plateau problem in Riemannian manifolds. *Matematichesky sbornik*, **V. 18, 3** (1972), pp. 487–527.

14. Fomenko, A.T., Minimal compacta in Riemannian manifolds and Reifenberg's conjecture. *Izvestiya AN SSSR*, **V. 6, 5** (1972), pp. 1037–1066.

15. Fomenko, A.T., Multidimensional Plateau problems on Riemannian manifolds and extraordinary homology and cohomology theories.
    Part 1. Trudy seminara po vektornomu i tensornomu analizu, Vyp. 18, Moscow University Press, Moscow, 1974, pp. 3–176.
    Part 2. Trudy seminara po vectornomu i tenzonomu analizu, Vyp. 18, Moscow University Press, Moscow 1978, pp. 4–93 (Russian).

16. Fomenko, A.T., On lower volume bounds for surfaces which are globally volume minimizing with respect to a cobordism constraint. *Izvestiya AN SSSR*, **V. 45, 1** (1981), pp. 187–212 (Russian).

17. Arnold, V., Varchenko, A.N. and Gusein-Zade, S.M., *Singularities of Differentiable Mappings*. Moscow, Nauka, 1982 (Russian).

18. Nitsche, J., *Vorlesunger über Minimalflächen*. Springer-Verlag, Berlin-Heidelburg-New-York, 1975.

19. Poston, T., The Plateau problem. An invitation to the whole of mathematics. Summar College on global analysis and its applications, 4 July–25 August 1972, International Centre for theoretical physics, Trieste, Italy.

20. Hurewicz, W. and Wallman, H., *Dimension Theory*. Princeton University Press, Princeton, N.J., 1941.

21. Borisovich, Yu.G., Bliznyakov, N.M., Izrailevich, Ya.A. and Fomenko, T.N., *Introduction to Topology*. Mir Publishers, Moscow, 1985.

22. Bombieri, E., De Giorgi, E. and Giusti, E., Minimal cones and the Bernshtein problem. *Invent. Math.*, **V. 7, 3** (1969), pp. 243–268.

23. Simons, J., Minimal varieties in Riemannian manifolds. *Ann. of Math.*, **V. 88, 2** (1968), pp. 62–105.

24. Integral Currents and Minimal Surfaces. *Sbornik perevodov*. Mir Publishers, Moscow, 1973 (Russian).

25. Hsiang, W. and Lawson, H., Minimal submanifolds of low cohomogeneity. *J. Diff. Geometry*, **V. 5, 1** (1971), pp. 1–28.

26. Lawson, H.B., The equivariant Plateau problem and interior regularity. *Trans. Amer. Math. Soc.*, **V. 173**, 1972, pp. 231–249.

27. Fomenko, A.T., *Variational Methods in Topology*. Nauka, Moscow, 1982 (Russian).

28. Dao Chông Thi, Multivarifolds and classical multidimensinal problems of Plateau. *Izvestiya AN SSSR*, **V. 44, 5** (1980), pp. 1031–1065 (Russian).

29. Plateau, J., *Statique Experimental et Théorique des Liquides Soumis aux Seules Forces Moléculaires*. Gauthier-Villars, Paris, 1873.

30. Lawson, H.B. and Osserman, R., Non-existence, non-uniqueness and irregularity of solutions to the minimal surface system. *Acta Math.*, **V. 139, 2** (1977), pp. 1–17.

31. Brothers, J., Invariance of solutions to invariant parametric variational problems. *Trans. Amer. Math. Soc.*, **V. 262, 1** (1980), pp. 159–180.

32. Rado, T., On the problem of Plateau. Springer-Verlag, Berlin, 1933.

33. Dao Chông Thi, Multidimensional variational problem in symmetric

spaces. *Funktsionalny analiz*, **V. 12, 1** (1978), pp. 72–73 (Russian).
34. Fomenko, A.T., Differential geometry and topology. Plenum Publ. Corporation, 1987, Ser. Contemporary Soviet Mathematics, Consultants Bureau, New York and London.
35. Suzuki, S., On homeomorphisms of 3-dimensional handlebody. *Canadian J. Math.*, **V. 29, 1** (1977), pp. 11–124.
36. Volodin, I.A. and Fomenko, A.T., On One Topological Algorithm *in* Tezisy dokladov VI Vsesoyuznoi topologicheskoy konferentsii. Minsk, Institut matematiki AN BSSR, 1977, p. 44 (Russian).
37. Eells, H. and Lemaire L., A report on harmonic maps. Bull. *London Math. Soc.*, **V. 10, 1** (1978), pp. 1–68.
38. Smith, R.T., Harmonic mappings of spheres. *Amer. J. of Math.*, **V. 97, 2** (1975), pp. 264–385.
39. Smith, R.T., The second variation formula for harmonic mappings. *Proc. Amer. Math. Soc.*, **V. 47, 1** (1975), pp. 229–236.
40. Volodin, I.A., Kuznetsov, V.E. and Fomenko, A.T., On the Problem of the Algorithmical Recognizability of the Standard Three-Dimensional Sphere. Moscow University Press, Moscow, **V. 24, 5** (1974), pp. 71–168.
41. Fomenko, A.T., Multidimensional Plateau problem on Riemannian manifolds. On the problem of the algorithmic recognizability of the standard three-dimensional sphere. *Proc. of the Intern. Congress of Math.*, Vancouver, **V. 1**, pp. 515–425.
42. Whitehead, J., On certain sets of elements in a free group. *Proc. of London Math. Soc.*, **V. 41** (1936), pp. 48–56.
43. Waldhausen, F., Heegard-Zerlegungen der 3-Sphare. Topology, **V. 7, 2** (1968), pp. 195–203.
44. Haken, W., Theorie der Normalflächen. *Acta Math.*, **V. 105** (1961), pp. 245–375.
45. Birman, J.S. and Hilden, H.M., Heegard splittings of branched coverings of S^2. *Trans. Amer. Math. Soc.*, **V. 213** (1975), pp. 315–352.
46. Homma, T., Ochaiai, M. and Takahashi, M., An algorithm for recognizing S^3 in 3-manifolds with Heegard splittings of genus two. *Osaka J. of Math.*, **V. 17** (1980), pp. 625–648.
47. Viro, O. Ya, and Kobelsky, V.L., The Volodin-Kuznetsov-Fomenko hypothesis on the Heegard diagrams of a three-dimensional sphere is not valid. *UMN*, **V. 32, 5** (1977), pp. 175–176 (in Russian).
48. Volodin, I.A. and Fomenko, A.T., Manifolds, knots and algorithms. *Sel. Math.*, **V. 3, 4** (1983/84), pp. 311–341.
49. Ochaiai, M., A counterexample to a conjecture of Whitehead and Volodin-Kuznetsov-Fomenko. *J. of Math. Soc.*, Japan, **V. 31** (1979), pp. 687–691.
50. Morikawa, O., A counterexample in the case of genus three. *Math. Sem. Notes of Kobe Univ.*, **V. 8, 2** (1980), pp. 295–298.
51. Smale, S., On the structure of manifolds. *Amer. J. of Math.*, **V. 84, 3** (1962), pp. 387–399.
52. Bogoyavlensky, O.I., On the exact function on manifolds. *Matematicheskie zametki*, **V. 8, 1** (1970), pp. 77–83.
53. Sharko, V. V., On Smooth Functions on Manifolds. Reprint, Institut matematiki AN UkSSR, Kiev, 1979.
54. Rourk, C. and Sanderson, P., *Introduction to Piecewise Linear Topology.* Berlin, Springer, 1982.

55. Kirby, R. and Scharleman, M., Eight faces of the Poincaré homological 3-sphere in Geometric Topology. New York, San Francisco, London, Acad. Press., 1979, pp. 113–136.
56. Bourbaki, N., Seminaire. Vols. 1981–1982. Exposé 585–590, Fevrier, Benjamin, New York and Amsterdam, 1982.
57. Shtanko, M.A., Menger problem solution in compactum class. DAN SSSR, V. **201, 6** (1971), pp. 1299–1302.
58. Chernavsky, A. V., On the k-stability of homeomorphisms and cell union. DAN SSSR, V. **180, 5** (1968), pp. 1045–1047.
59. Novikov, S.P., Multi-valued functions and functionals. A Morse theory analog. DAN SSSR, V. **260, 1** (1981), pp. 31–35.
60. Novikov, S.P., Hamilton formalism and the multi-valued analog of Morse theory, UMN, V. **37, 5** (1982), pp. 3–49.
61. Novikov, S.P. and Schmeltser, I., Periodic solutions of the Kirchhoff equations of free motion of a solid body in the ideal fluid and extended Lyusternik-Schnirelmann-Morse Theory. 1. Funktsionalny analiz, V. **15, 3** (1981), pp. 54–66. 2. Funksionalny analiz, V. **15, 4** (1981) (Russian).
62. Freedman, M. H., The topology of four-dimensional manifolds. *J. of Diff. Geom.*, V. **17, 3** (1982), pp. 357–453.
63. Sharko, V. V., Exact Morse functions on 1-connected manifolds with 1-connected boundary. UMN, V. **36, 5** (1981).
64. Yau, S.T., Kohn-Rossi cohomology and its application to the complex Plateau problem. *Ann of Math.*, V. **113, 1** (1981), pp. 67–110.
65. Federer, H. and Fleming, W.H., Normal and integral currents. *Ann. Math.*, V. **72, 2** (1960), pp. 458–520.
66. Osserman, R., Global properties of classical minimal surfaces in E^3 and E^n *Duke Math. J.*, V. **32, 4** (1965), pp. 565–573.
67. Fomenko, T.N., On the efficient construction of the retraction of some spaces onto the sphere *in* Analysis on Manifolds and Differential Equations. Voronezh University Press, Voronezh, 1986, pp. 164–173 (Russian).
68. Fomenko, A. T., Symmetries of soap films. *Comp. and Maths. with Appls.*, V. **12B, 3–4** (1986), pp. 825–834. See also Symmetry (Unifying Human Understanding), István Hargittai, Pergamon Press, International Series in Modern Applied Mathematica and Computer Science, V. **10**, 1986.
69. Dao Chông Thi and Fomenko, A.T., *Minimal Surfaces and the Plateau Problem*. Nauka, Moscow, 1987 (Russian).
70. Birman, S. and Hilden, M., The homeomorphism problem for S^3. *Bull. of Amer. Math. Soc.*, V. **79, 5** (1973), pp. 1006–1010.
71. Tuzhilin, A.A. and Fomenko, A.T., Multivalued mappings, minimal surfaces and soap films. *Vestnik MGU*, Moscow University Press, Moscow, Ser. 1, *Matematika i mekhanika* 3 (1986), pp. 3–12 (Russian).
72. Shklyanko, I.V., Minimal geodesics in symmetry spaces *in* Geometry, Differential Equations and Mechanics. Moscow University Press, Moscow, 1986, pp. 159–161 (Russian).
73. Dao Chong Thi, Multivarifolds and problems of minimizing functionals of multidimensional volume type. *DAN SSSR*, V. **276, 5** (1984), pp. 1042–1045.
74. Dao Chông Thi, Isoperimetric inequalities for multivarifolds. *Izvestiya AN SSSR*, V. **48, 4** (1984), pp. 1031–1065 (Russian).
75. Le Hong Van, New examples of globally minimal surfaces *in* Geometry, Differential Equations and Mechanics. Moscow University Press, Moscow, 1986, pp. 102–105 (Russian).

76. Le Hong Van, The increase of a two-dimensional minimal surface. *UMN*, **V. 40, 3** (1985), pp. 209–210 (Russian).
77. Le Hong Van, Minimal surfaces and the calibration forms in symmetry spaces. *Trudy seminara po vektornomu analizu, Vyp. 22*, Moscow University Press, Moscow, 1985, pp. 107–118 (Russian).
78. Le Hong Van and Fomenko, A.T., Lagrangian manifolds and the Maslov index in the minimal surface theory. *DAN SSSR*, **V. 299, 1** (1988), pp. 42–45.
79. Le Hong Van and Fomenko, A.T., The test for minimality of Lagrangian subsurfaces in Kähler manifolds. *Matematicheskie zametki*, **V. 4**, 1987, pp. 559–571 (Russian).
80. Le Hong Van, Minimal surfaces in homogeneous spaces. *Izvestiya AN SSSR*, **V. 51, 2** (1988), pp. 408–423 (Russian).
81. Tuzhilin, A. A., On bifurcation of certain two-dimensional minimal surfaces under two-parametric contour variation *in* Geometry, Differential Equations and Mechanics. Moscow University Press, Moscow, 1986, pp. 140–145 (Russian).
82. Ivanov, A.O., Globally minimal symmetric surfaces in Euclidean space *in* Geometry, Differential Equations and Mechanics. Moscow Press University, Moscow, 1986, pp. 69–71 (Russian).
83. Balinskaya, I.S., Orbit volumes of smooth actions of Lie groups *in* Geometry, Differential Equations and Mechanics. Moscow University Press, Moscow, 1986, pp. 49–51.
84. Balinskaya, I.S., Minimal cones of the adjoint action of classical Lie groups. *UMN*, **V. 41, 6**(252) (1986), pp. 165–166 (in Russian).
85. Balinskaya, I.S., Minimal cones of the adjoint action of classical Lie groups *in* Tezisy Bakinskoy mezhdunarodnoy topologicheskoy konferentsii, Institut matematiki i mekhaniki AN AzSSR, Baku, **V. 2**, 1987, p. 32 (Russian).
86. Tyrin, A.V., On a property of local minima absence of the multidimensional Dirichlet functional. *UMN*, **V. 39**, Vyp. 2., 1984, pp. 193–194 (Russian).
87. Tyrin, A.V., Critical points of the multidimensional Dirichlet functional. *Matematichesky Sbornik*, **V. 124, 1** (1984), pp. 146–158 (Russian).
88. Tyrin, A.V., The regularity of Riemannian manifolds mapping minimizing the multidimensional Dirichlet functional. *Matematichesky Sbornik*, **V. 132, 3** (1987), pp. 401–419 (Russian).
89. Tyrin, A.V., The regularity of harmonic mappings as a derived relation of being unstable. *DAN SSR*, 1987 (in print) (Russian).
90. Pluzhnikov, A.I., On harmonic mappings of Riemannian surfaces and fibered manifolds. *Matematichesky Sbornik*, **V. 113, 2** (1980), pp. 340–347.
91. Pluzhnikov, A.I., Some properties of harmonic mappings in the case of sphere and Lie groups. *DAN SSSR*, **V. 268, 6** (1983), pp. 1300–1302.
92. Pluzhnikov, A.I., Certain geometric properties of harmonic mappings. *Trudy seminara po vektornomu i tenzornomu analizu*, **V. 22**, Moscow University Press, Moscow, 1985, pp. 132–147 (Russian).
93. Pluzhnikov, A.I., On the Dirichlet functional minima. *DAN SSSR*, **V. 290, 2** (1986), pp. 289–293.
94. Pluzhnikov, A.I., Topological criterion of energy functional global minimum inattainability *in* Analysis on Manifolds and Differential Equations. Voronezh University Press, Voronezh, 1986, pp. 149–155 (Russian).

95. Pluzhnikov, A.I., Minimization problem for energy functional. Reprint No. 5584-84, *VINITI*, Moscow, 1984 (Russian).

96. Tuzhilin, A.A., Indices of two-dimensional minimal surfaces *in* Geometry and Singularity Theory in Non-Linear Equations. Voronezh University Press, Voronezh, 1987, pp. 170–176 (Russian).

97. Tuzhilin, A.A., About the indices of minimal surfaces *in* Tezisy Bakinskoy mezhdunarodnoy topologicheskoy konferentsii, Baku, Institut matematiki i mekhaniki AN Az.SSR, V. 2, 1987, p. 301 (Russian).

98. Beeson, M. and Tromba, A., The cusp catastrophe of Thm in the bifurcation of minimal surfaces. *Manusc. Mathem.*, V. 46, pp. 272–307.

99. Böhme, R., Hildebrandt, S. and Tausch, E., The two-dimensional analogue of the catenary. *Pacific J. Math.* V, 88, 2 (1980), pp. 247–278.

100. Calabi, E., Minimal immersions of surfaces in Euclidean spheres. *J. Diff. Geom.*, V. 1, 2 (1967), pp. 111–125.

101. Freed, D. and Uhlenbeck, K., Instantons and Four-Manifolds. Springer, N.Y., 1984.

102. Eells, J. and Wood, J., Maps of minimum energy. *J. London Math. Soc.*, *Ser.* 2, V. 23, 2 (1981), pp. 303–310.

103. Grüter, M., Hildebrandt, S. and Nitsche, J.C., On the boundary behavior of minimal surfaces with a free boundary, which are not minima of area. *Manusc. Math.*, V. 35, 3 (1981), pp. 387–410.

104. Gulliver, R. and Spruck, J., On embedded minimal surfaces. *Ann. Math.*, V. 103, 2 (1976), pp. 331–347.

105. Harvey, R. and Lawson, H. B., On boundaries of complex analytic varieties. *Ann. of Math.*, V. 102, 2 (1975), pp. 223–290.

106. Harvey, B. and Lawson, H.B., Calibrated geometries. *Acta Math.*, V. 148, 1982, pp. 47–157.

107. Hildebrandt, S., Boundary behavior of minimal surfaces. *Arch. R. Mech.*, V. 35, 1 (1969), pp. 47–82.

108. Meeks, W.H. and Yau, S.T., Topology of three-dimensional manifolds and the embedding problems in minimal surface theory. *Ann. of Math.*, V. 112, 3 (1980), pp. 441–484.

109. Morgan, F., A smooth curve in R^3, bounding a continuum of minimal surfaces. *Arch. R. Mech.*, V. 71, 2 (1–81), pp. 193–197.

110. Sacks, J. and Uhlenbeck, K., Minimal immersions of closed Riemann surfaces. *Trans. Amer. Math. Soc.*, V. 271, 2 (1982), pp. 639–652.

111. Schoen R. and Uhlenbeck, K., Regularity of minimizing harmonic maps into the sphere. *Invent. Math.*, V. 78, 1 (1984), pp. 89–100.

112. White, B., Homotopy classes in Sobolev spaces and energy minimizing maps. *Bull. Amer. Math. Soc. New Ser.*, V. 13, 2 (1985), pp. 166–168.

113. Tomi, F. and Tromba, A.J., Existence theorems for minimal surfaces of non-zero genus spanning a contour. Memoirs Amer. Math. Soc., V. 71, 382 (1988).

114. Borisovich, A.Yu., Reducing the problem of bifurcation of minimal surfaces to the operator equations and search bifurcations of the catenoid, helicoid, Scherk and Enneper surfaces. *Uspekhi Mat. Nauk*, Vyp. 5(251), V. 41, 1986, pp. 165–166.

115. Tromba, A., Degree theory on oriented infinite-dimensional varieties and the Morse number of minimal surfaces spanning a curve in R^n. Part I. *Trans. AMS*, V. 290, 1 (1985), pp. 385–413.

116. Tromba, A., Degree theory on oriented infinite-dimensional varieties and the Morse number of minimal surfaces spanning a curve in R^n. Part II. *Manuscr. Math.*, V. 48, 1–3 (1984), pp. 139–161.

117. Smale, S., On the Morse index theorem. *J. Math. Mech.*, **V. 14, 6** (1965), pp. 1049–1055.
118. Beeson, M., Some results on finiteness in Plateau's problem. *I. Math. Z.*, **V. 175, 2** (1980), pp. 103–123.
119. Bohme, R. and Tromba, A., The index theorem for classical minimal surfaces. *Ann. Math.*, **V. 113, 3** (1981), pp. 447–499.
120. Barbosa, J. and Do Carmo, M., On the size of a stable minimal surface in R^3. *Amer. J. Math.*, **V. 98, 2** (1976), pp. 515–528.
121. Do Carmo, M. and Peng, C., Stable complete minimal surfaces in R^3 are planes. *Bull. Amer. Math. Soc.*, **V. 1, 6** (1979), pp. 903–906.
122. Do Carmo, M. and Dajczer, M., Rotation hypersurfaces in spaces of constant curvature. *Trans. Amer. Math. Soc.*, **V. 277, 2** (1983), pp. 685–709.
123. Borisovich, A. Yu., The Plateau Operator and Bifurcation of Two-Dimensional Minimal Surfaces. *Global analysis and Mathematical Physics.* Voronezh, 1987, pp. 141–154 (Russian).
124. Rassias, T. M., Foundations of global non-linear analysis. *Teubner-text zür Math.*, Leipzig, Bd. 86, 1986.
125. Mori, H., Minimal surfaces of revolution in H^3 and their stability properties. *Indiana Math. J.*, **V. 30**, 1987, pp. 787–794.
126. Barbosa, J. L. and Do Carmo, M., Stability of minimal surfaces and eigenvalue of the Laplacian. *Math. Z.*, Bd. 173, 1, s. 13–28.
127. Osserman, R., Minimal surfaces. *Uspekhi Mat. Nauk*, **V. 22, 4** (1967), pp. 55–136 (Russian).
128. Fischer-Colbrie, D., On complete minimal surfaces with finite Morse index in three-manifolds. *Invent. Math.*, **V. 82, 1** (1985), pp. 121–132.
129. White, B., Homotopy classes in Sobolev spaces and the existence of energy minimizing maps. *Acta Math.*, **V. 160, 1–2** (1988), pp. 1–18.

INDEX

Printed and bound by CPI Group (UK) Ltd, Croydon, CR0 4YY

24/10/2024

01778279-0003